Biofluid Mechanics

Second Edition

Biofluid Mechanics

Second Edition

Jagannath Mazumdar

School of Electrical and Electronic Engineering
The University of Adelaide, Australia

in collaboration with

S. Sircar, A. Saha and K. Wong

 World Scientific

NEW JERSEY · LONDON · SINGAPORE · BEIJING · SHANGHAI · HONG KONG · TAIPEI · CHENNAI · TOKYO

Published by

World Scientific Publishing Co. Pte. Ltd.

5 Toh Tuck Link, Singapore 596224

USA office: 27 Warren Street, Suite 401-402, Hackensack, NJ 07601

UK office: 57 Shelton Street, Covent Garden, London WC2H 9HE

Library of Congress Cataloging-in-Publication Data
Mazumdar, J. (Jagannath)
   Biofluid mechanics / by Jagannath Mazumdar (The University of Adelaide, Australia). --
2nd edition.
     pages cm
   Includes bibliographical references and index.
   ISBN 978-9814713979 (hardcover : alk. paper)
   1. Hemodynamics. 2. Rheology (Biology). 3. Body fluid flow. I. Title.
   QP105.M378 2015
   612.1'181--dc23

                       2015028011

British Library Cataloguing-in-Publication Data
A catalogue record for this book is available from the British Library.

Printed in Singapore

Dedicated in loving memory to
My Departed Parents
and to
My Beloved Guru
Bhagawan Sri Sathya Sai Baba

CONTENTS

Chapter 4   MODELS OF BIOFLUID FLOWS

Chapter 5   NON-NEWTONIAN FLUIDS

PREFACE TO THE SECOND EDITION

The second and enlarged edition of *BIOFLUID MECHANICS* is due to the enthusiastic response to the first edition by the editorial board members of the World Scientific, Singapore. This present edition incorporates many of the suggestions received from my colleagues and students advising me to include materials related to Tissue Biomechanics, a new emerging field.

However, the general plan, as described in the preface to the first edition published in 1992, remains unchanged, except for the addition of two new chapters (9 and 10) on Tissue Engineering and Cellular Engineering and some new materials on computational biofluid mechanics that have been included in Chap. 8. These two chapters have been written in collaboration with my colleague, Dr. Sarthok Sircar, of the University of Adelaide and an ex-student, Professor Asit Saha of the Central State University, Ohio, USA. I am extremely grateful to them for preparing these two chapters. The new addition in Chap. 8 is mainly by one of my ex-students Dr. Kelvin Wong for which I am extremely grateful to him. Although I cannot express my individual thanks to all readers of the first edition for their valuable comments, their communications stimulated me during the past so many years to think of improvements. I hope that the new edition will make for easier reading and teaching from the book.

As mentioned in the previous edition, this volume can be regarded as a complement to my previous book *An Introduction to Mathematical Physiology and Biology* published by Cambridge University Press, First Edition in 1989, Second Edition in 1999. Readers who are familiar with this book, will find the present book a natural continuation of the previous one.

While preparing this new edition, I have received assistance from many people. My heartfelt thanks go to all of them. In particular, I would like to thank Sarthok Sircar for meticulously typesetting the manuscript and making a thorough proof reading and for his unremitting efforts to get this second edition to the highest editorial level possible. I wish to express my appreciation to my wife, Maya, for her personal care and encouragement during the preparation of the revised manuscript. I also wish to acknowledge the School of Electrical and Electronic Engineering, University of Adelaide, where the final preparation of the book is completed.

A few inaccuracies which slipped our notice in the first edition have also been corrected in this edition. In spite of great effort on the part of many proofreaders, mistakes may appear in this edition. I would be most grateful if these are brought to my attention so that they could be attended in the future.

J. MAZUMDAR
ADELAIDE, AUSTRALIA

PREFACE TO THE FIRST EDITION

For more than a decade I have been involved in studying various aspects of biofluid mechanics. During that period, I have been fortunate to have had a number of excellent collaborators as graduate students and research workers from whom I learned a lot on this subject. In preparing this monograph I have borrowed substantially from the work of these collaborators and I have frequently quoted from their publications.

The present book can be regarded as a complement to my previous book *An Introduction to Mathematical Physiology and Biology* published by Cambridge University Press in 1989. Readers who are familiar with my previous book, will find the present book a natural continuation of the previous one.

The book gives a systematic account of a number of topics commonly discussed under physiological and biological fluid dynamics. It begins with an introduction to elementary fluid mechanics as taught at undergraduate level. The basic equations of motion for both viscous and inviscid fluids are given in detail. This is followed by the mechanics of flow in the circulatory system. Circulatory mechanics is a wide subject and in the present book only systemic and pulmonary circulations are explained. In Chap. 3 flow properties of blood are presented. A brief mention of some diseases related to obstruction of blood flow is made. Chapter 4 discusses some well-known models of biofluid flows. Applications to duct and pipe flow are also given. Non-Newtonian fluid flow is discussed in Chap. 5. Some models of other flows are included in Chap. 6, i.e., flows in capillary blood vessels, in kidneys and the peristaltic flows. The fluid mechanics of heart valves, natural and prosthetic, are discussed in Chap. 7. The last chapter deals with the computational methods in biofluid mechanics.

In preparing an account of biofluid mechanics, however, there may be difficulties in finding the right balance between biological and physiological fluid motion on one side and mathematical modelling on the other side. The balance has been given in the present book.

Some of the topics presented in the monograph are old and well-known, but some are relatively new. Although I wanted to include a much wider selection of topics, I found that this was not possible because of the degree of thoroughness that would have been required. I therefore wish to apologize in advance for this limitation.

There are many people whose help I must acknowledge. I wish to mention especially my former students Kym Thalassoudis, Lai-Leng Soh and Keng Cheng Ang whose research have given me motivation in writing this book. I am also thankful to

Miss Danielle Hopkins for meticulously typesetting the manuscript using LaTeX and to Mrs Jenny Laing for her excellent care and cooperation in drawing the figures for the book. Finally, I wish to express my gratitude to my wife, Maya, for the unlimited personal encouragement, patience and support I received during the course of writing this monograph and without which it would never have been finished.

In spite of great care and precautions, I am sure that many mistakes and defects remain in the book. I would be most grateful if these are brought to my attention so that they could be attended in the future.

J. MAZUMDAR

ADELAIDE, AUSTRALIA

Chapter 1

INTRODUCTION

1.1 A Basic Introduction

Biological Science has long been advanced from purely descriptive to analytical science. Many analytical methods of physical science have been used successively in the study of biological science. Biofluid mechanics is the study of certain class of biological problems from fluid mechanics points of view. Biofluid mechanics does not involve any new development of the general principles of fluid mechanics but it does involve some new applications of the method of fluid mechanics. Complex movements of fluids in the biological system demand for their analysis professional fluid mechanics skills. The most common fluid mechanics problem in the biological system is the flow of blood. In 1840, the French physician Poiseuille was interested in the flow of blood and conducted a study of flow in capillaries. It is the well-known Poiseuille flow in fluid mechanics, but we know that ordinary Poiseuille flow does not represent the actual blood flow in a biological system. Many biofluid mechanics problems are concerned with classical fluid mechanics, but also its modern aspects such as rheology, chemical reactions, electrothermal effects etc. Approaching the problem from a macroscopic (continuum approach) point of view by assuming that any small volume element of the fluid is so large that it contains a very large number of molecules and also assuming that the fluid is a continuous medium, the dependent variables describing the fluid motion are as follows:

$$\text{pressure } p = p\,(x, y, z, t),$$
$$\text{density } \rho = \rho\,(x, y, z, t),$$
$$\text{velocity } \mathbf{q} = \mathbf{q}\,(x, y, z, t) = (u, v, w).$$

The number of basic variables in fluid mechanics is five: three velocity components and two thermodynamic properties. The flow field is completely determined once we specify the velocity vector **q** and two thermodynamic properties as a function of space and time. Hence we need five independent equations. These are usually the three components of the equation of motion, a continuity equation and an energy equation. We will discuss this in the next section.

If the density ρ is constant (does not vary with pressure) throughout then the flow is said to be *incompressible*. In fluid mechanics, it is customary to treat liquids as incompressible fluids. In an incompressible fluid the energy equation is not needed since density is taken as known and only pressure together with velocity need to be found in order to completely describe the flow of fluids.

In general, we divide fluids into two classes — viscous and non-viscous (inviscid). A brief description of viscosity will now be given for the benefit of the reader.

1.1.1 *Viscosity*

Newton defined the viscosity of a fluid as a lack of slipperiness between the layers of the fluid and, of course, in doing so he implied that there was such a thing as a "layer of fluid" or "laminae" of the fluid, and the viscosity arises as a result of rubbing one lamina upon the other.

Consider two laminae which are in contact with one another (see Fig. 1.1).

Fig. 1.1: Newtonian concept of viscosity.

Suppose some force F parallel to the x-axis acts and produces relative motion between the lamina, i.e., the top lamina moves with velocity dv relative to the bottom lamina. Hence, there is a rate of change of velocity with distance in the y direction, i.e., there exists a velocity gradient $\frac{dv}{dy}$. It is hypothesized that

$$F \propto A \frac{dv}{dy} \qquad (1.1)$$

where A is the area of contact between the laminae. The proportionality constant is then defined to be the viscosity of the fluid, and is usually denoted by μ, i.e.,

$$F = \mu A \frac{dv}{dy}. \qquad (1.2)$$

The dimensions of the viscosity are given by

$$[\mu] = \frac{[\text{force}]}{[\text{area}]\frac{[\text{velocity}]}{[\text{length}]}} = \frac{MLT^{-2}}{L^2 \frac{LT^{-1}}{L}} = \frac{M}{LT}. \qquad (1.3)$$

It is customary to classify fluids as *Ideal* or *non-Viscous* (μ=0), *Newtonian* (μ is constant), or *non-Newtonian* (μ is variable). Actually, no fluid is ever really ideal, but some fluids, at least in certain regions of flow under certain circumstances, approach ideal conditions very closely.

1.1.2 *Laminar and Turbulent Flow*

The terms laminar flow and purely viscous flow are used synonymously to mean a fluid flow which flows in laminae, as opposed to turbulent flow in which the velocity components have random fluctuations imposed upon their mean values. For a given fluid, as the velocity increases the flow changes from laminar to turbulent, passing through a transition region. Both laminar and turbulent flow occur in nature, but turbulent flow seems to be a more natural state of affairs.

1.1.3 *Compressible and Incompressible Flow*

It is common practice to divide fluids into two groups — liquids and gases. Gases are compressible and their density changes readily with temperature and pressure. Liquids, on the other hand, are rather difficult to compress and for all practical purposes may be considered as incompressible.

From above it can be seen that there are many different properties that a fluid may exhibit. However, the theoretical development of fluid mechanics can primarily be broken down into two classes:

(i) Inviscid fluid mechanics (i.e., non-viscous or viscosity can be neglected)

(ii) Viscous fluid mechanics

In the next section, we give the basic equations of fluid mechanics for application in subsequent sections.

1.2 Basic Equations of Fluid Mechanics

As mentioned earlier, there are five basic variables in fluid flow, three components of velocity and two thermodynamic properties. Hence there are five basic equations which describe the flow; the continuity equation, three components of the momentum equation, and the energy equation. In general, the energy equation becomes uncoupled in *incompressible* flow since the density is constant. In turbulent flow the situation is somewhat more complex and a closed set of equations cannot generally be developed.

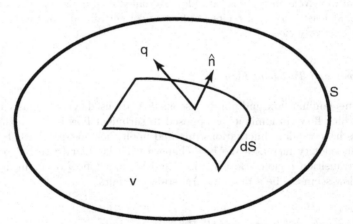

Fig. 1.2: Volume V enclosed by surface S.

1.2.1 *Continuity Equation*

Consider a fixed volume V in space in our flow field with closed surface S (see Fig. 1.2). Assuming there are no sources or sinks in V and applying the Principle of Conservation of Mass, that is,

Time rate of change of total mass in V
= Rate of inflow (or outflow) of mass across S

we get:

$$\frac{\partial}{\partial t} \int_V \rho \, dV = \int_S -\rho \, \mathbf{q}.\hat{\mathbf{n}} \, dS \tag{1.4}$$

where ρ = density at a point in V at time t
$\quad\;\; \mathbf{q}$ = velocity at a point in V at time t
$\quad\;\; \hat{\mathbf{n}}$ = outward unit normal to the surface at a point.

The fact that the density of a fluid can vary from point to point makes the fluid compressible.

Using Gauss' Divergence Theorem, that is

$$\int_S \mathbf{F}.\hat{\mathbf{n}} \, dS = \int_V div\mathbf{F} \, dV \tag{1.5}$$

Eq. (1.4) becomes

$$\int_V \frac{\partial \rho}{\partial t} \, dV = \int_V -div(\rho \, \mathbf{q}) \, dV,$$

i.e., $\quad\; \int_V \left[\frac{\partial \rho}{\partial t} + div(\rho \, \mathbf{q}) \right] dV = 0. \tag{1.6}$

As the choice of V is arbitrary,

$$\frac{\partial \rho}{\partial t} + div(\rho \, \mathbf{q}) = 0$$

or, $\quad \frac{\partial \rho}{\partial t} + \mathbf{q}.\nabla\rho + \rho \, div\mathbf{q} = 0, \tag{1.7}$

i.e., $\quad \frac{D\rho}{Dt} + \rho \, div\mathbf{q} = 0$

where the material derivative is

$$\frac{D}{Dt} \equiv \frac{\partial}{\partial t} + \mathbf{q}.\nabla. \tag{1.8}$$

Equation (1.7) is called the *Conservation of Mass equation* or the *Continuity equation*. It is widely applicable as no special assumption has been made so far as to what type of fluid can be considered.

In the case of an incompressible fluid, we have

$$\frac{D\rho}{Dt} = 0. \tag{1.9}$$

Hence, the continuity equation becomes

$$div\mathbf{q} = 0, \tag{1.10}$$

i.e.,

$$\frac{\partial u}{\partial x} + \frac{\partial v}{\partial y} + \frac{\partial w}{\partial z} = 0 \tag{1.11}$$

where u, v and w denote velocity components in the x, y and z directions respectively.

1.2.2 *Equations of Motion*

In a fluid, forces can be divided into two types, namely, external forces (viz. gravitational, body forces etc.) and internal forces.

External forces are forces acting on a local portion of fluid due to some agency outside the fluid. It is expressed as a force per unit mass, $\mathbf{F}$. So in a volume δV of the fluid, we have mass $\rho \delta V$ with associated external or body force $\mathbf{F} \rho \delta V$.

Internal forces are forces due to interaction between neighbouring fluid particles. These are expressed as forces per unit area or *stresses* and are exerted across the boundary surface between elements.

Let S be a surface between the fluid considered and the 'external' fluid (can be a fluid of the same type) and dS be an element of S with unit normal $\hat{\mathbf{n}}$. Also let $\mathbf{T}$ be the stress vector (force per unit area) giving the force between the internal and external fluids. Then the internal forces acting between sections of fluid for the surface element dS is

$$(\mathbf{T}.\hat{\mathbf{n}})\, dS\, \hat{\mathbf{n}} = \mathbf{T}^n dS \quad \text{(say)}. \tag{1.12}$$

In a three-dimensional case, a full description of these forces requires vectors in three mutually perpendicular directions, that is, we need a nine-component entity — a tensor $\sigma_{ij}, (i, j = 1, 2, 3)$. (σ_{ij}) is called the *stress tensor* and σ_{ij} corresponds to the force component per unit area. The first subscript denotes the face on which the stress acts, and the second subscript denotes the direction of stress. The face is indicated as the plane perpendicular to the axis of the subscript. For instance, the 3rd face is perpendicular to the z or x_3 axis (see Fig. 1.3).

In tensor notation,

$$\mathbf{T}^n dS = \mathbf{T}^j n_j dS$$
$$\text{or,} \quad T_i^n dS = T_i^j n_j dS$$
$$= \sigma_{ij} n_j\, dS$$
$$= \sigma_{ij}\, dS_j. \tag{1.13}$$

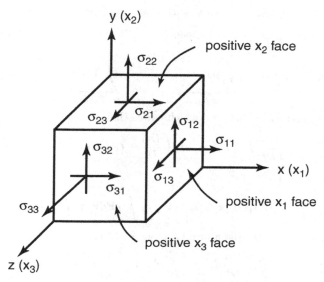

Fig. 1.3: Stresses at a point in space. The positive faces are shown; the opposite ones are negative faces.

Let us now consider a viscous fluid. The inviscid case can be obtained simply by putting the viscosity, μ, to be zero.

Assuming symmetry of the stress tensor, that is,

$$\sigma_{ij} = \sigma_{ji} \tag{1.14}$$

and using the Newtonian relation for the viscous part

$$\text{stress} = \text{viscosity} \times \text{rate of strain},$$

we obtain

$$\sigma_{ij} = \underbrace{-p\,\delta_{ij}}_{\text{inviscid part}} + \underbrace{\mu\left(\frac{\partial u_i}{\partial x_j} + \frac{\partial u_j}{\partial x_i}\right)}_{\text{viscous part}} \tag{1.15}$$

where u_i = velocity of the fluid flow in the x_i-direction
$\quad\; p$ = pressure acting on the surface (hydrostatic pressure)
$\quad\; \mu$ = coefficient of viscosity
$\quad\; \delta_{ij} = \begin{cases} 1 & \text{if } i = j \\ 0 & \text{otherwise.} \end{cases}$

Consider a general fluid volume V with bounding surface S as in Fig. 1.2. Since

Time rate of change of momentum = Net force acting on fluid,

we have

$$\frac{D}{Dt} \int_V \rho\, \mathbf{q}\, dV = \underbrace{\int_V \rho\, \mathbf{F}\, dV}_{net\, body\, force} + \underbrace{\int_S \mathbf{T}^n dS}_{net\, surface\, force\, due\, to\, external\, fluid} \qquad (1.16)$$

Using Reynolds Transport theorem, which states that

$$\frac{D}{Dt} \int_V G\, dV = \int_V \left(\frac{DG}{Dt} + G\, div\mathbf{q} \right) dV \qquad (1.17)$$

we obtain from Eq. (1.16)

$$\int_V \left[\frac{D}{Dt}(\rho\mathbf{q}) + \rho\mathbf{q}\, div\mathbf{q} \right] dV = \int_V \rho\mathbf{F}\, dV + \int_S \mathbf{T}^n dS.$$

For an incompressible fluid, the above becomes

$$\int_V \rho \frac{D\mathbf{q}}{Dt}\, dV = \int_V \rho\mathbf{F}\, dV + \int_S \mathbf{T}^n\, dS \qquad (1.18)$$

or in indicial notation,

$$\int_V \rho \frac{Dq_i}{Dt}\, dV = \int_V \rho F_i\, dV + \int_S T_i^n\, dS. \qquad (1.19)$$

Substituting Eq. (1.13) into Eq. (1.19),

$$\int_V \rho \frac{Dq_i}{Dt}\, dV = \int_V \rho F_i\, dV + \int_S \sigma_{ij}\, dS_j. \qquad (1.20)$$

Applying Gauss' Divergence theorem, this becomes

$$\int_V \rho \frac{Dq_i}{Dt}\, dV = \int_V \rho F_i\, dV + \int_V \frac{\partial \sigma_{ij}}{\partial x_j}\, dV. \qquad (1.21)$$

As the choice of V is arbitrary, we get

$$\rho \frac{Dq_i}{Dt} = \rho F_i + \frac{\partial \sigma_{ij}}{\partial x_j}. \qquad (1.22)$$

From Eq. (1.15),

$$\begin{aligned}
\frac{\partial \sigma_{ij}}{\partial x_j} &= -\frac{\partial p}{\partial x_j}\delta_{ij} + \mu\frac{\partial^2 u_i}{\partial x_j \partial x_j} + \mu\frac{\partial^2 u_j}{\partial x_j \partial x_i} \\
&= -\frac{\partial p}{\partial x_i} + \mu\nabla^2 u_i + \mu\frac{\partial}{\partial x_i}\left(\frac{\partial u_j}{\partial x_j}\right) \\
&= -\frac{\partial p}{\partial x_i} + \mu\nabla^2 u_i + \mu\frac{\partial}{\partial x_i}(div\mathbf{q}). \qquad (1.23)
\end{aligned}$$

In the case of an incompressible fluid, $div\mathbf{q} = 0$. Thus

$$\frac{\partial \sigma_{ij}}{\partial x_j} = -\nabla p|_i + \mu \nabla^2 u_i. \tag{1.24}$$

Substituting this into Eq. (1.22) we get in vector form

$$\rho \frac{D\mathbf{q}}{Dt} = \rho \mathbf{F} - \nabla p + \mu \nabla^2 \mathbf{q}. \tag{1.25}$$

This is known as the *Navier-Stokes equation* for a viscous, incompressible fluid. In the Cartesian coordinate system this reduces to

$$\rho \left(\frac{\partial u}{\partial t} + u \frac{\partial u}{\partial x} + v \frac{\partial u}{\partial y} + w \frac{\partial u}{\partial z} \right) = \rho F_x - \frac{\partial p}{\partial x} + \mu \nabla^2 u \tag{1.26}$$

$$\rho \left(\frac{\partial v}{\partial t} + u \frac{\partial v}{\partial x} + v \frac{\partial v}{\partial y} + w \frac{\partial v}{\partial z} \right) = \rho F_y - \frac{\partial p}{\partial y} + \mu \nabla^2 v \tag{1.27}$$

$$\rho \left(\frac{\partial w}{\partial t} + u \frac{\partial w}{\partial x} + v \frac{\partial w}{\partial y} + w \frac{\partial w}{\partial z} \right) = \rho F_z - \frac{\partial p}{\partial z} + \mu \nabla^2 w. \tag{1.28}$$

If, however, the flow is frictionless then there are no shear stresses and the normal stresses are just the pressure which is isotropic, i.e.,

$$\sigma_{ij} = -p \, \delta_{ij}. \tag{1.29}$$

In this case the equations of motion become

$$\rho \frac{D\mathbf{q}}{Dt} = \rho \mathbf{F} - \nabla p. \tag{1.30}$$

These equations are known as the *Euler equations* for frictionless flow. If the flow is incompressible, the density is constant and the three Euler equations (1.30) plus the continuity equation (1.10) constitute four equations with four unknowns u, v, w, and p.

If we compare the complete Navier-Stokes equations (1.25) with the Euler equations (1.30), we notice that the differences are the additional terms involving the viscosity. However, if the fluid is incompressible, even if the viscosity is not zero, the viscous terms in Eq. (1.25) are negligible if derivatives of the form $\frac{\partial u_i}{\partial x_j}$ are small, and the Navier-Stokes equations reduce to the Euler equations. Thus, the Navier-Stokes equations reduce to the Euler equations if the viscosity is negligible or the derivatives of velocities are negligible. Indeed, the last conclusion is very important in fluid mechanics because for many fluid flows it is assumed that there are two regions: one close to a solid boundary where the viscous terms are important and thus the Navier-Stokes equations are considered, and a region away from the boundary where

Euler's equations hold good approximately. The flows of fluid in these two regions are referred to as "Boundary Layer Flow" and "Potential Flow". We will discuss these in more detail in subsequent chapters.

1.3 Simplification of Basic Equations

The continuity equation and the equations of motion can be simplified in the following cases:

(i)   When there are *no external body forces*, i.e., when $F_x = F_y = F_z = 0$, or when the external forces form a conservative system. If the external forces form a conservative system, then they can be derived as the gradient of a scalar potential function Ω such that

$$\rho\, F_x = -\frac{\partial \Omega}{\partial x}, \quad \rho\, F_y = -\frac{\partial \Omega}{\partial y}, \quad \rho\, F_z = -\frac{\partial \Omega}{\partial z}. \tag{1.31}$$

In this case the terms $(\rho\, \mathbf{F} - \nabla p)$ appearing in Eq. (1.25) become

$$\rho \mathbf{F} - \nabla p = -\nabla(p + \Omega) \tag{1.32}$$

so that the pressure is effectively replaced by $p + \Omega$.

(ii) When the flow is *steady*, i.e., when the flow does not vary with time so that u, v, w and p are functions of x, y, z only and $\frac{\partial u}{\partial t}, \frac{\partial v}{\partial t}, \frac{\partial w}{\partial t}$ and $\frac{\partial p}{\partial t}$ are all zero. If the flow is such that the velocity components and thermodynamic properties at every point in space change in time, the flow is called *unsteady*.

(iii) When the flow is *two-dimensional*, i.e., when the flow is the same in all planes parallel to say, the xy-plane. In this case we assume that $w = 0$ and there is no variation with respect to z. In such a case we may think of a volume of fluid of unit thickness moving parallel to the xy-plane. Thus, the continuity equation (1.11) and the Navier-Stokes equations (1.26)-(1.27) in the absence of external body forces take the form

$$\frac{\partial u}{\partial x} + \frac{\partial v}{\partial y} = 0 \tag{1.33}$$

$$\rho\left(\frac{\partial u}{\partial t} + u\frac{\partial u}{\partial x} + v\frac{\partial u}{\partial y}\right) = -\frac{\partial p}{\partial x} + \mu\left(\frac{\partial^2 u}{\partial x^2} + \frac{\partial^2 u}{\partial y^2}\right) \tag{1.34}$$

$$\rho\left(\frac{\partial v}{\partial t} + u\frac{\partial v}{\partial x} + v\frac{\partial v}{\partial y}\right) = -\frac{\partial p}{\partial y} + \mu\left(\frac{\partial^2 v}{\partial x^2} + \frac{\partial^2 v}{\partial y^2}\right). \tag{1.35}$$

We note that Eq. (1.33) is identically satisfied if we assume a function $\psi(x, y)$ such that

$$u = \frac{\partial \psi}{\partial y}, \quad v = -\frac{\partial \psi}{\partial x}. \tag{1.36}$$

The function $\psi(x, y)$ is called the *Stream function* since $\psi(x, y) = const$ gives us a streamline for the flow. Clearly, from Eq. (1.36) we note that the directional derivative of ψ in any direction gives the component of the velocity in a direction making an angle $-90°$ with the direction of the derivative. Since the velocity vector $\mathbf{q}$ is tangent to a streamline, it has no component normal to the streamline, hence $\frac{\partial \psi}{\partial s} = 0$ where s is measured along the streamline, which implies that $\psi = const$ along each streamline. Obviously, the value of ψ will vary from streamline to streamline. If we define dn as the element in a direction normal to ds and if the angle between ds and dn is $90°$ then we have

$$\frac{\partial \psi}{\partial n} = q \quad \text{and} \quad d\psi = q \, dn. \tag{1.37}$$

Thus, the increment $d\psi$ in ψ from one streamline to the next equals the amount of fluid passing between these two streamlines per unit time. It therefore follows that the amount of fluid crossing any curve AB (Fig. 1.4) in unit time is $\psi_B - \psi_A$, which is the value of ψ at B minus the value of ψ at A. Using the stream function ψ, Eqs. (1.34) and (1.35) can be combined and after eliminating p between them transform into

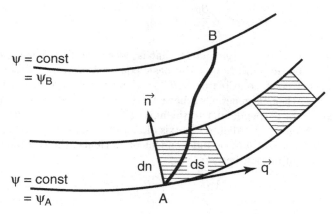

Fig. 1.4: The streamlines and the stream function ψ.

$$\frac{\partial}{\partial t}\nabla^2\psi + \frac{\partial\psi}{\partial y}\frac{\partial}{\partial x}\nabla^2\psi - \frac{\partial\psi}{\partial x}\frac{\partial}{\partial y}\nabla^2\psi = \nu\nabla^4\psi \qquad (1.38)$$

where ν is the kinematic viscosity μ/ρ. If further we introduce the *vorticity* of this two-dimensional flow as

$$\omega = \frac{1}{2}\left(\frac{\partial v}{\partial x} - \frac{\partial u}{\partial y}\right) = -\frac{1}{2}\nabla^2\psi \qquad (1.39)$$

then Eq. (1.38) reduces to

$$\frac{\partial\omega}{\partial t} + \frac{\partial\psi}{\partial y}\frac{\partial\omega}{\partial x} - \frac{\partial\psi}{\partial x}\frac{\partial\omega}{\partial y} = \nu\nabla^2\omega. \qquad (1.40)$$

(iv) When the flow is *one-dimensional*, i.e., if there is only one non-zero velocity component, say u, then the continuity equation reduces to

$$\frac{\partial\rho}{\partial t} + \frac{\partial}{\partial x}(\rho u) = 0 \qquad (1.41)$$

or, if the fluid is incompressible

$$\frac{\partial u}{\partial x} = 0 \qquad (1.42)$$

and the Navier-Stokes equations become a single equation

$$\rho\frac{\partial u}{\partial t} = -\frac{\partial p}{\partial x} + \mu\left(\frac{\partial^2 u}{\partial y^2} + \frac{\partial^2 u}{\partial z^2}\right) + \rho F_x. \qquad (1.43)$$

(v) When the flow is *axially symmetric*, that is, when the flow is symmetric about an axis, further simplification of the basic equations can be carried out. If we use cylindrical polar coordinates (r, θ, z), where the axis of symmetry is along the axis of z, then in general we have three components of velocity; v_r along the radius vector perpendicular to the axis, v_θ perpendicular to the axis and the radius vector, and v_z parallel to the axis of z. In the axi-symmetric case, we take $v_\theta = 0$, and v_r, v_z and p to be independent of θ. Hence, the continuity equation and equations of motion become

$$\frac{1}{r}\frac{\partial}{\partial r}(rv_r) + \frac{\partial}{\partial z}v_z = 0, \qquad (1.44)$$

$$\rho\left(\frac{\partial v_r}{\partial t} + v_r \frac{\partial v_r}{\partial r} + v_z \frac{\partial v_r}{\partial z}\right) = -\frac{\partial p}{\partial r} + \mu\left(\frac{\partial^2 v_r}{\partial r^2} + \frac{1}{r}\frac{\partial v_r}{\partial r} - \frac{v_r}{r^2} + \frac{\partial^2 v_r}{\partial z^2}\right), \quad (1.45)$$

$$\rho\left(\frac{\partial v_z}{\partial t} + v_r \frac{\partial v_z}{\partial r} + v_z \frac{\partial v_z}{\partial z}\right) = -\frac{\partial p}{\partial z} + \mu\left(\frac{\partial^2 v_z}{\partial r^2} + \frac{1}{r}\frac{\partial v_z}{\partial r} + \frac{\partial^2 v_z}{\partial z^2}\right). \quad (1.46)$$

We can further simplify these equations if we introduce the stream function ψ defined by

$$v_z = \frac{1}{r}\frac{\partial \psi}{\partial r}, \quad v_r = -\frac{1}{r}\frac{\partial \psi}{\partial z}. \quad (1.47)$$

With this form of ψ, Eqs. (1.45) and (1.46) can be combined to yield the following equation in ψ

$$\frac{\partial}{\partial t}(\nabla_1^2 \psi) - \frac{1}{r}\frac{\partial(\psi, \nabla_1^2 \psi)}{\partial(r,z)} - \frac{2}{r^2}\frac{\partial \psi}{\partial z}\nabla_1^2 \psi = \nu\nabla_1^4 \psi \quad (1.48)$$

where

$$\nabla_1^2 \equiv \frac{\partial^2}{\partial r^2} - \frac{1}{r}\frac{\partial}{\partial r} + \frac{\partial^2}{\partial z^2}, \quad \nabla_1^4 \equiv \nabla_1^2(\nabla_1^2)$$

and $\quad \dfrac{\partial(\psi, \nabla_1^2 \psi)}{\partial(r,z)} = \dfrac{\partial \psi}{\partial r}\dfrac{\partial(\nabla_1^2 \psi)}{\partial z} - \dfrac{\partial \psi}{\partial z}\dfrac{\partial(\nabla_1^2 \psi)}{\partial r}. \quad (1.49)$

1.4 Initial and Boundary Conditions

The above equations can be solved subject to initial and boundary conditions. Initial conditions give the motion of the fluid at time $t = 0$ while boundary conditions are prescribed on the surfaces with which the fluid may be in contact. Contact is preserved between the fluid and the rigid surfaces and at a solid-fluid interface the fluid must not penetrate the solid if it is impermeable to the fluid. Therefore the normal velocity of the contact surface and that of the fluid are the same, i.e., the relative velocity component of the fluid normal to the solid surface must vanish. In the case of a fixed rigid boundary, the boundary conditions are provided by the no-slip condition according to which both tangential and normal components of the fluid velocity vanish at all points of the surfaces of the stationary bodies with which the fluid may be in contact. However, if the rigid boundary is in motion, the relative velocity of the fluid is tangential to the surface; this is also true for two fluids in contact along a common surface.

1.5 Dimensional Analysis in Fluid Mechanics

In fluid mechanics a great many problems may be solved using dimensional analysis. The relevant dimensionless parameters in fluid mechanical situations may be combined into independent dimensionless groups characterising the flow. These dimensionless parameters, or π's as they are called, may be obtained by the method of dimensional analysis. Dimensional analysis, or the *Buckingham's π-Theorem* as it is called (Buckingham [14]), is a method of finding the relevant dimensionless parameters without knowing the relevant differential equations. In fact, the essence of the theorem states that if a physical process is governed by a dimensionally homogeneous relation involving n-dimensional parameters then there exists an equivalent relation involving a smaller number, $(n - k)$, of dimensionless parameters, called π-variables. The reduction, k, is usually equal to but never more than, the number of fundamental dimensions involved in the physical process. However, the method requires a knowledge of all the relevant variables in any physical problem. For example, as explained earlier, for the flow of viscous incompressible fluid only the continuity equations and the Navier-Stokes equations are necessary to describe the flow. There are four independent variables u, v, w and p — the three velocity components and the pressure. In addition, there are the fluid properties, density ρ, viscosity μ, and the gravitational potential. In fact, in the Navier-Stokes equations (1.26)-(1.28), the terms on the left-hand sides represent the *inertial forces* while the three terms on the right-hand side of each equation represent respectively the *body forces, pressure forces and viscous forces*. Let us put these equations in dimensionless form. Choose a characteristic velocity V and a characteristic length L. For example, if we were to consider the flow in a tube, V could be the mean flow speed and L, the tube diameter. On the other hand, if we were to consider the external flow over a cylinder, L could be the diameter and V the free stream velocity. Having chosen these characteristic quantities, we introduce the following dimensionless variables:

$$x' = x/L, \ \ y' = y/L, \ \ z' = z/L,$$

$$u' = u/V, \ \ v' = v/V, \ \ w' = w/V, \tag{1.50}$$

$$p' = p/\rho V^2, \ \ t' = Vt/L$$

and the parameter

$$R_e = VL\rho/\mu. \tag{1.51}$$

In terms of these variables, Eqs. (1.26)-(1.28) in the absence of body forces, can be written in dimensionless or normalised form:

$$\frac{\partial u'}{\partial t'} + u'\frac{\partial u'}{\partial x'} + v'\frac{\partial u'}{\partial y'} + w'\frac{\partial u'}{\partial z'} = -\frac{\partial p'}{\partial x'} + \frac{1}{R_e}\left(\frac{\partial^2 u'}{\partial x'^2} + \frac{\partial^2 u'}{\partial y'^2} + \frac{\partial^2 u'}{\partial z'^2}\right) \tag{1.52}$$

and two similar equations in v' and w'. The equation of continuity (1.11) can also be put in dimensionless form

$$\frac{\partial u'}{\partial x'} + \frac{\partial v'}{\partial y'} + \frac{\partial w'}{\partial z'} = 0. \tag{1.53}$$

The parameter R_e defined in Eq. (1.51) is known as the *Reynolds number* which is a dimensionless quantity, probably the most important dimensionless parameter in fluid mechanics. It is clear that only one physical parameter, the Reynolds number R_e, enters into the basic equations of fluid flow. Thus, the flows over two geometrically similar bodies (same shape, but different in size) will be identical if the Reynolds numbers for the two bodies are the same.

Consider now the physical significance of the Reynolds number. It will be shown that the Reynolds number is a measure of the ratio of inertial forces to the viscous forces. In a flow the inertial forces, in terms of characteristic velocity and characteristic length are of the order $\rho V^2/L$ and the viscous forces are of the order $\mu V/L^2$. The ratio of these two forces is

$$\frac{\text{Inertial forces}}{\text{Viscous forces}} = \frac{\rho V L}{\mu} = \text{Reynolds number}. \tag{1.54}$$

Clearly, when the Reynolds number is small, $R_e \ll 1$, the viscous forces dominate. A small Reynolds number will occur when the viscosity is large or when either the characteristic length or characteristic velocity is small. A flow with small Reynolds number is sometimes called *slow* motion. On the other hand, when the Reynolds number is large, $R_e \gg 1$, the inertial forces are large compared to viscous forces, and the inertial forces tend to diffuse the fluid particles, causing rapid mixing in the fluid, which is a characteristic of *turbulence*.

In a straight pipe, turbulence occurs when R_e exceeds a critical value of about 2000. In the case of blood flow, laminar flow continues to occur at Reynolds numbers as high as 10,000. In fact, in many biological situations high Reynolds number laminar flow is more common than turbulent flow.

Now that we have introduced the properties of fluids and have some idea of how a fluid behaves under certain circumstances, we can put our study of fluid mechanics into perspective by developing its application in biological systems. This then will be the essence of the following chapters.

Chapter 2

CIRCULATORY BIOFLUID MECHANICS

2.1 General Introduction

A unicellular organism can simply absorb nutrients, required to maintain its life, from its immediate environment, and just as easily excrete its waste products into the environment. Life is maintained so long as the environment is large enough to provide ample supply of nutrients, and is not polluted.

However, in the human body, which is a multi-cellular body, each cell acts as if it were a unicellular organism — it takes nutrients from its immediate environment and excretes waste-products into the same environment. This immediate environment is enclosed within the body. It is not large, and hence there is a need for a system which can transport materials (both nutrients and waste products) around the body. Such a system does exist and is termed the *circulatory system.*

The circulatory system of a living body functions as a means of transporting and distributing essential substances (e.g. oxygen, hormones) to the tissues and removing by-products (e.g. carbon dioxide, water) of metabolism. Apart from serving as an important transport system, the circulatory system is also essential in such homeostatic mechanism as regulation of body temperature, humoral communication throughout the body, and adjustment of oxygen and nutrient supply in different physiological situations.

Hence there must be adequate circulation at all times to the important organs of the body — brain, heart and lungs. Should circulation fail or malfunction, various diseases and even death could occur. Thus, the importance of the circulatory system cannot be overemphasised.

In this chapter, a description of the circulatory system is given and several diseases related to abnormalities, malfunction or simply failure of the circulatory system are discussed (see Hellums and Brown [38]).

2.2 The Circulatory System

2.2.1 *Introduction*

The circulatory system is made up of the heart and blood vessels. It performs one main function — to transport. This system is responsible for delivering oxygen and nutrients needed for metabolic processes to the tissues, carrying waste products from cellular metabolism to the kidneys and other excretory organs for elimination, and circulating electrolytes and hormones needed to regulate body function. It also regulates the body temperature by moving core heat to the periphery, where it can be dissipated into the external environment. In addition, it plays the vital role of transporting various immune substances that contribute to the body's defence mechanisms.

In other words, the circulatory system basically consists of a central pump (the heart) and the peripheral blood vessels (arteries, veins and capillaries). The transport medium, blood, is circulated throughout the body in a network as depicted in Fig. 2.1. The peripheral blood vessels form a sequential continuum insofar as blood flow is concerned, i.e., blood flows successively from one type of the vessel into another. The circuit is complete as can be seen from Fig. 2.1.

2.2.2 *Systemic and Pulmonary Circulations*

The circulatory system can be divided into two sections: the systemic and the pulmonary circulations (see Fig. 2.2).

The systemic circulation supplies blood to all of the body's tissues except the lungs. As it needs to pump blood to distant parts of the body, often against gravity, the systemic circulation functions as a high-pressure system with arterial pressure up to 100 mm Hg. The lungs are supplied by the pulmonary circulation, which provides a gas exchange function. It is located in the chest in close proximity to the heart, which propels blood through it. Thus, the pulmonary circulation functions as a low-pressure system.

In Table 2.1, the first horizontal row shows the mean pressure (mm Hg) in the various blood vessels for the systemic circulation while the second horizontal row shows that for the pulmonary circulation.

The arteries carry blood pumped by the heart under pressure to the tissues and various organs of the body, supplying them with the essential substances via smaller vessels called arterioles and capillaries. After useful substances are absorbed into the tissues, blood is returned to the heart by the low-pressure smaller veins (venules) and the larger veins (vena cava). The structural properties of arteries and veins are explained in Sec. 3.5.

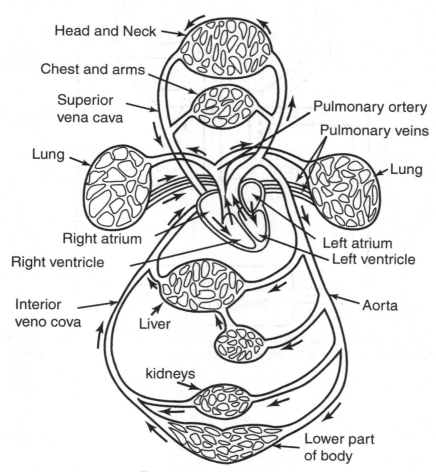

Fig. 2.1: Circulation of blood.

Fig. 2.2: The systemic and pulmonary circulation are arranged in series. Arrows indicate direction of blood flow (full arrows: oxygenated blood, dotted arrows: deoxygenated blood).

Table 2.1: Mean pressures (mm Hg) in blood vessels

Aorta	Arteries	Arterioles	Capillaries	Venules	Veins	Vena Cava	RA	RV
100	90 →	75 →	45 →	25 →	10 →	5 →	5 →	25
↑								↓
LV	LA	Veins	Capillaries	Arterioles		Arteries		Pulmonary artery
100	← 5	← 5	← 10	← 15	←	20	←	25

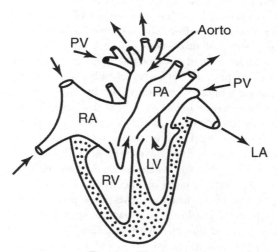

Fig. 2.3: The heart.

In essence, the efficient functioning of the circulatory system is essential to the well-being of the human body. For the system to function smoothly, a smooth flow through the distributory system is necessary.

2.2.3 *The Circulation in the Heart*

The distributory system comprises of arteries and arterioles responsible for sending blood to the various organs of the body. Fine capillaries make up the diffusing system and act as an exchange system. The collecting system consists of veins which collect blood depleted of oxygen and full of products of metabolic processes of the cardiovascular system.

Each division of the circulation includes a pump, a distributory system, a diffusing system and a collection system. The heart is the pump that propels blood through both circulations. It is divided into a right heart, which delivers blood to the pulmonary circulation and a left heart, which delivers blood to the systemic circulation (see Fig. 2.2). Each side of the heart is further divided into two chambers, an atrium and a ventricle. The ventricles are the main pumping chambers of the heart. The atria act as collection chambers for various blood returning to the heart and as auxiliary pumps for the ventricles. Thus, the heart is a four-chambered structure that generates the force to make the blood flow. There has not been a pump or engine devised that can even equal the long-term performance of the heart. Consider, for instance, an average man with a heart-beat rate of 70 beats per minute lived for 70 years. His heart would have completed about 2.6×10^{10} beats before it stopped.

As mentioned earlier, the heart actually consists of two parts — the right heart and the left heart, connected in series: the output of one part becomes the input of the other. Blood that returns from the systemic circulation (see Fig. 2.2) which supplies relevant and essential substances to the body, enters the right side of the heart. The blood is, under normal circumstances, deoxygenated (having given up its oxygen content to various parts of the body) when it enters the right heart. It is then pumped through the pulmonary circulation (the lungs) where it "collects" oxygen from and releases carbon dioxide to the air sacs of the lungs. Now with a fresh supply of oxygen, it returns to the left heart, which upon the next pump, sends blood into the systemic circulation again. Unidirectional flow of blood within the circulatory system is achieved by means of four valves strategically situated in the inflow and outflow tracts of each ventricle; this is also assisted functionally by suitable pressure gradients from arterial to venous parts of the circuit.

The "pumping" action of the heart is a two-phase process: the contraction phase when blood is forced out of the heart is known as *systole*; the relaxation, or filling phase when blood is received into the heart is termed *diastole*. During systole, the pressures in the aorta (main artery from left heart) and the pulmonary artery rise to 120 mm Hg and 25 mm Hg respectively. At the end of the systole, the valves close, and there is an abrupt drop in pressure inside the heart (*intraventricular pressure*). In order that blood flow is not abruptly disturbed or interfered by this phasic change, the arteries and other vessels must have some characteristic structure.

Different vessels of the circulatory system have quite different structural, and hence physiological properties. In this section, we shall concern ourselves mainly with structure and morphology of arteries and arterioles, making only brief comments on capillaries, veins and other vessels.

The basic morphologic arrangement of the wall structure of a typical artery is shown in Fig. 2.4. Note that there are in fact two layers of elastic membranes on the wall of the artery. The innermost layer of the artery is the endothelium. Depending on the composition of the constituents, the relative diameter and the distance from the heart, the arteries can be classified as elastic arteries, muscular arteries or arterioles.

Elastic arteries are large and have within their walls prominent series of sheets composed of a sclerprotein, elastin. The wall of a large elastic artery, e.g. the aorta, is thick. In proportion to the cross-sectional area of the lumen, (the "space" enclosed by the arterial wall) however, the aortic wall is relatively thinner than the wall of the smaller muscular arteries. The walls of some large thick-walled arteries have their own blood supplies.

Muscular arteries are blood vessels intermediate between the large elastic arteries and the smaller arterioles. Being the most numerous arterial structure to be found in the body, muscular arteries form the major distributing arterial system. The walls of this class of artery include a thick layer of smooth muscle fibres, with collagenous tissues. Elastic connective tissues are also present. These arterial components are continuous with surrounding loose connective tissues, and thus their movements permit the vessel to change its diameter readily. In mathematical or physical terms, the vessels are not rigid but elastic. Longitudinally, however, the arteries are under a state of tension and this property is evident from the fact that the vessels retract when severed (Fung [28]).

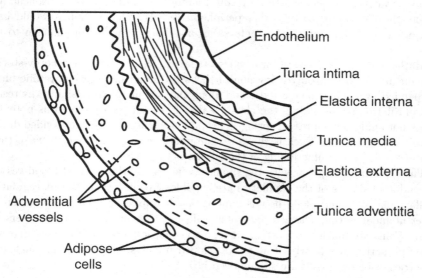

Fig. 2.4: Cross-section of a typical artery to illustrate the several layers of tissue that are present in such vessels.

The arterioles are the smallest arteries with diameter 0.3 mm or less. Only larger arteries have an internal elastic membrane, and although external elastic membranes are also found in smaller arteries, they are absent in arterioles. However, loose connective fibres including elastic and collagenous fibres are present in arterioles.

The anatomic transition from one type of artery to the next may be abrupt or progressive but a smooth flow of blood is generally maintained in a normal body. One important point to note here is the fact that the entire cardiovascular system has an inner endothelial lining, which makes it possible for it to be a smooth and continuous, uninterrupted channel in which for blood to flow.

However, the contractions of the heart are intermittent, or pulsatile and so blood is ejected in spurts and not a smooth stream. If the walls of the arteries and arterioles were rigid, then the flow of blood in the capillaries would likewise be pulsatile instead of steady. Since the exchange of substances between blood and interstitial fluid takes place through the thin walls of the capillaries, it is necessary to have a smooth, continuous flow in the capillaries. Fortunately, the large arteries near the heart are normally elastic and hence distensible. Consequently, they damp the pulsatile systolic output of the ventricles. During contraction, the ejection of blood distends the aorta and its large arterial branches. With the closure of the aortic valves, the elastic arteries recoil, thereby sustaining the pressure head and making the blood flow to the other peripheral vessels steadier than it would have been otherwise. The artery thus acts as an ancilliary pumping mechanism to the heart.

Mathematically, the elastic properties of the arteries provide an "energy-storing mechanism" — so that only a fraction of the work performed during systolic phase is expended in moving the ejected blood along the peripheral vessels, and the rest is used in dilating or distending the elastic walls. Energy is hence stored as potential energy not unlike a compressed spring. This potential energy is then expended during diastole (see Fig. 2.5). Such is the very characteristic of arteries which make them so vital in the proper functioning of the circulatory system.

The capillaries, together with the smallest venules, are the smallest blood vessels. The endothelial cells of the capillaries are capable of contraction. Hence, regulating capillary diameter and blood flow is unnecessary in capillaries. Veins are present in higher abundance than the arteries, their walls are thinner and their diameters are larger. Veins are much less elastic and tend to collapse more readily. Muscular and elastic properties are poorly, if at all, developed in veins. The basic morphological properties of blood vessels are summarised in Fig. 2.6.

2.3 Diseases Related to Circulation

Let us examine briefly some of the diseases related to the circulatory system. Vascular diseases are undoubtedly important. They weaken vessels and make them vulnerable to rupture. Such diseases also narrow and occlude the vascular lumens, thereby cutting down the blood supply to the tissues and organs. Common arterial

Fig. 2.5: Stages of contraction and dilation of aortic walls, illustrating the "energy-storing" mechanism with analogy to that of a spring.

Fig. 2.6: Summary of various properties of different blood vessels.

diseases include Arteriosclerosis, Vasculitis, Raynaud's Disease, and Aneurysms. Venous diseases like Varicose Veins and venous thromboses are also common.

Arteriosclerosis is the generic term for three patterns of vascular diseases, all of which cause thickening and inelasticity of arteries. The dominant pattern is *atherosclerosis*. It is the most common ailment of the Western populations. The other patterns are Mönckeberg's medial calcific sclerosis affecting the muscular arteries, and arteriolosclerosis which involves smaller arteries and arterioles.

As a person gets older, their arteries may have large amounts of cholesterol deposits, or calcifications may occur in the media of muscular arteries (in individuals over 50 years of age). This changes the structure of the arteries, and the most important changes are the lowering of elasticity of arterial walls, and the thickening of walls (hence reducing lumens). We have seen the importance of the elastic property of arteries in maintaining a smooth blood flow. With inelastic arteries and reduced lumens, coronary diseases like *myocardial infarction* (*"heart attack"*) may occur. "Infarction" is a general term used to describe the situation when death of tissues occurs as a result of oxygen deficiency. A particular organ may fail depending on the degree of damage. In myocardial infarction, the organ is the heart.

Vasculitis is characterised by the inflammation of an artery (or vein) caused by the spread of a contiguous inflammatory process, such as a bacterial infection. Abnormal dilatations of arteries or veins are called *Aneurysm* — occurring more frequently in arteries, especially in the aorta. They are developed whenever there is a marked weakening of the wall of a vessel. Any vessel may be affected by a wide variety of disorders, including congenital defects, local infections (mycotic aneurysms), trauma or systemic diseases that weaken arterial walls.

It is not the purpose of this section to dwell on the details of the various diseases, which forms too large a subject by itself. Nevertheless, one should by now realise the extensive relevance and importance of the blood vessels and blood flow in pathology. Variations from the normal characteristics of blood flow vessels or blood itself can lead to some serious clinical problems. We will discuss this aspect in greater detail in the next chapter.

Chapter 3

BLOOD RHEOLOGY: PROPERTIES OF FLOWING BLOOD

3.1 General Introduction

Having discussed the structural properties of the arteries, we now turn our attention to the transport medium, namely, the blood.

Physiologically, amongst other functions, blood serves as the major bulk transportation and distribution system for materials within the body — transport of respiratory gases, nutrients, waste products of metabolism, etc. Because blood is a suspension of particles in an aqueous solution, it is not a homogeneous fluid and has some unusual properties.

The previous chapter confirms the importance of a knowledge of blood vessels and blood flow characteristics, as it may help in detecting, if not designing a treatment for, the various diseases. For instance, the disease, *Diabetes Mellitus* is related to the unusual viscous properties of blood. It has been studied and found that the viscosity of diabetic blood may be raised or lowered depending on the shear rate of blood. Studies have also been made to determine the correlation between viscosity factors of blood and chronic anxiety states. The patients, apart from having anxiety disorders were supposed to be in normal health and yet blood viscosity was higher in them.

3.2 Blood Composition

Firstly, let us look at the constituents of blood.

Table 3.1: Blood constituents (5×10^6 particles/mm^3).

Cell Elements	Relative Proportions
Red cells (erythrocytes)	600
White cells (leucocytes)	1
Platelets (thrombocytes)	30

Plasma	Weight Fraction
Water	0.91
Proteins	0.07
Inorganic solutes	0.01
Other organic substances	0.01

Table 3.2: Normal range of values for the cellular elements concentration in the human blood.

	Average cell concentration cells/ml	Approximate normal range cells/ml	Characteristic length μm	Percentage of total volume cells %
Total W.B.C.	9000	4000-11000	10	5
Neutrophils	5400	3000-6000		
Eosinophils	270	150-300		
Basophils	60	0-100		
Lymphocytes	2730	1500-4000		
Monocytes	540	300-600		
Total R.B.C.			1.9×8.0	91
Female	4.8×10^6			
Male	5.4×10^6			
Platelets	3000000	150000-400000	3	4

Blood, a complex body fluid, is a liquid tissue consisting of several types of formed elements (or cells) suspended in an aqueous fluid matrix (the plasma). The plasma itself contains a complex spectrum of organic molecules. It is made up of Red Cells (RBC or erythrocytes), White Cells (WBC or leucocytes) and Platelets (thrombocytes). Note that RBC constitute about 95%. Plasma is an aqueous solution of electrolytes and organic substances, mainly proteins. These proteins include fibrinogen, globulins, albumin, beta lipoprotein and lipalbumin. The first three make up approximately 5%, 45% and 50% of plasma protein respectively. The latter two are present in very small proportions. Fibrinogen, though not present in high concentration, is the largest of the proteins while albumin is the smallest. Tables 3.1 and 3.2 show the relative proportions of cell elements and the concentrations of plasma.

The WBC play a major role in fighting diseases. WBC are relatively fewer in number compared to the RBC, and so their contribution to the circulatory characteristic parameters is insignificant in general, though in some pathological cases, this may not be true. The platelets are very numerous, but they are also very small in size. Because of their small size, they are usually neglected or considered unimportant where blood flow discussion is concerned.

The RBC consist of a very flexible membrane enclosing concentrated solution haemoglobin. The viscosity of haemoglobin is 5 times that of blood, and this makes it possible for RBC to deform. This property enables RBC, of average diameter 8 μm not only to pass through capillaries (5 μm in diameter), but also through the endothelial wall. RBC play a major role in determining the mechanical properties of blood. It is hence important to know the volume concentration of RBC. The quantity, haematocrit is used to represent this volume concentration, being defined simply as a percentage volume concentration. It ranges from 42 to 45 in normal blood composition.

The specific gravity of red cells is about 1.06, while that of plasma is 1.03, consequently if blood is allowed to stand in a container, the red cells will settle out of suspension. They do so at a definite rate called the *Erythrocyte Sedimentation Rate* or ESR for short. The mathematical theory of this process is very simple and can be studied on the basis of the theory of fluid mechanics. In illness, the ESR always increases abruptly. This is due to the formation of red cell aggregation.

3.3 Structure of Blood

Blood cells form a continuous structure when at rest (see Fig. 3.1). Sedimentation occurs when the structure breaks under its own weight and when a finite stress is applied to the whole blood, the structure breaks. The stress needed to cause this breakage is called the yield stress. After this breakage, the structure becomes a suspension of a cluster of cell aggregates in plasma. These aggregates are in turn formed from smaller units of cells called rouleaux [9].

There exists a dynamic equilibrium between the size of aggregates or rouleaux and the stress applied. Under sufficiently high stress, aggregates and rouleaux are reduced to individual cells, which will not be reduced any further by further increases in stress. Thus, the shear stress-shear strain rate relationship becomes a straight line at sufficiently high shear stresses (see Fig. 3.2).

Fig. 3.1: A suspension of cells at rest is a continuous structure formed from groups of cell aggregates made up of units of cells called rouleaux or flocs.

Fig. 3.2: The relationship between shear stress and shear strain rate for blood plasma.

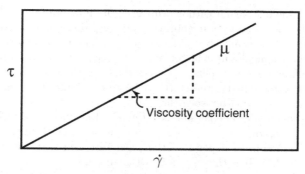

Fig. 3.3: Dynamic equilibrium between cell aggregate and shear rate normal plasma; (hematocrit $\simeq$ 45%).

In fact, the shear stress-shear strain rate relationship of plasma is normally considered linear as shown in Fig. 3.3. This is, of course, the property of a Newtonian fluid. Thus, plasma is considered to be Newtonian. However, there are several studies which describe plasma as non-Newtonian. It can thus be concluded that plasma may have transitory non-Newtonian character at low shear rates.

3.4 Flow Properties of Blood

3.4.1 *Viscosity of Blood*

Blood properties are dominated by the properties of RBC as about 95% of blood cells are RBC (except in cases like leukaemia) (see Table 3.1). A human red cell is a disc-shaped element having a mean diameter of 7.2 μm and a thickness of about 2.2 μm at the thickest part, that is, near the circumference, and about 1 μm at the center (see Fig. 3.4).

Fig. 3.4: Diagram showing the dimensions of the red cell.

It has an average area of 120 μm^2 and a volume of 85 μm^3. The biconcave shape serves to increase the surface area available for diffusion of oxygen and allows the cell to change in volume and shape without rupturing its membrane.

The red cell is bounded by a membrane made of protein in association with lipid and steroid materials, a semi-permeable membrane. The most important element in the red cell is the haemoglobin, which transports oxygen to the tissues.

The volume fraction of red cells in blood is known as the hematocrit, H. However, this value is normally a little greater than the true volume fraction, ϕ, as a small volume of plasma is trapped between the cells.

Blood is neither homogeneous nor Newtonian. Plasma, when isolated, may be Newtonian with a viscosity of about 1.2 times that of water. However, the viscosity for whole blood is found to depend on the shear rate and is measured as apparent viscosity (see Fig. 3.3). It is thus a non-Newtonian fluid.

For values of ϕ up to 0.05 a constant shear stress-shear strain rate relationship typical of a Newtonian fluid is observed. The viscosity of blood for $\phi < 0.05$ is expressed with reasonable accuracy by Einstein's equation for spheres in suspension (Jeffery [46])

$$\mu_s = \mu_p \left(\frac{1}{1 - \alpha\phi} \right) \tag{3.1}$$

where μ_s = suspension viscosity
μ_p = plasma viscosity
α = shape factor (2.5 for spheres)
ϕ = cell volume fraction.

In theory, the constant α should vary with the shape of the particles, but a value of 2.5 appears to apply to red cells when $\phi < 0.05$.

However, Eq. (3.1) is only good under the assumption that the particles act individually and do not interact significantly. As ϕ increases from 0.05, this assumption is not met as interaction between particles becomes more significant. The properties associated with interaction, for example, cell elasticity and aggregation, now influence viscosity, which departs from Eq. (3.1).

Later, the use of Eq. (3.1) with a variable α is proposed, where

$$\alpha = 0.076 \exp \left[2.49\phi + \frac{1107}{T} e^{(-1.69\phi)} \right] \tag{3.2}$$

and T = absolute temperature ($^\circ K$).

This is found to agree with measurements of apparent viscosity of red cell suspensions for ϕ up to 0.6. Figure 3.5 shows a plot of α against ϕ for $T = 25^\circ$ and 37°.

3.4.2 *Yield Stress of Blood*

A number of non-linear expressions have been suggested as a means to characterise blood viscosity and yield stress of blood. For example, Casson [16] suggested an expression

$$\sqrt{\tau} = K\sqrt{\gamma} + \sqrt{c} \qquad (3.3)$$

whereas Herschel-Bulkley suggested

$$\tau = b\gamma^s + c \qquad (3.4)$$

where K = Casson viscosity,
c = yield stress of blood, and
b, s = constants.

As mentioned before, blood requires a finite stress to disrupt the aggregates before it begins to flow. For hematocrits $H > 5.8\%$, this yield stress is found to be given by

$$\tau_0^{\frac{1}{3}} = \frac{A(H - H_m)}{100} \qquad (3.5)$$

where A = $(0.008 \pm 0.002\,\mathrm{dyne/cm^2})^{\frac{1}{3}}$
H = normal hematocrit
H_m = hematocrit below which there is no yield stress.

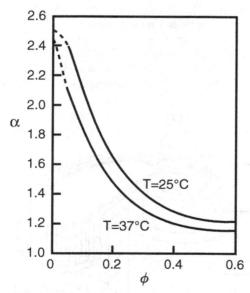

Fig. 3.5: Plot of α against ϕ for ϕ up to 0.6.

Taking $H = 45\%$ and $H_m = 5\%$, the yield stress of normal human blood works out to be between 0.01 and 0.06 dyne/cm^2.

Other than the yield stress τ_0, shear stress τ and apparent viscosity μ are also dependent on the hematocrit. On top of this, the apparent viscosity is observed to depend on capillary radius for capillaries of diameter less than 300 μm. This phenomenon is known as the Fahraeus-Lindqvist effect.

The yield stress is determined from the cell to cell contact (see Fig. 3.1). A number of ingenious methods have been used for the measurement of yield stress. However, there is an inherent difficulty in the measurement of yield stress. The measuring instruments interact with the red cell structure and little is known about this interaction. Hence, it is not apparent whether measurements of blood yield stress are in fact measurements of the structure of blood or the structure-instrument interface. We will discuss blood viscosity again in the next chapter after we discuss models of blood flow.

3.5 Blood Vessel Structure

Except for the capillaries, all blood vessels have walls composed of three layers separated by an elastic lamina (see Fig. 3.6). The *tunica externa*, or *tunica adventitia*, is the outermost covering of the vessel. This layer is composed of fibrous and connective tissue that serves to support the vessel. This layer is separated by the external elastic lamina from the middle layer, or the *tunica media*, which is largely a smooth muscle that constricts and relaxes in order to control the diameter of the vessel. The *tunica intima*, or inner layer, has an elastic layer called the internal elastic lamina, that joins the media and a thin layer of epithelial cells that lie adjacent to the blood. The epithelial layer provides a smooth and slippery inner surface for the vessel. This smooth inner lining, as long as it remains intact, prevents blood clotting.

Fig. 3.6: Medium-sized artery and veins showing the relative thickness of the three layers.

3.5.1 *Arteries and Arterioles*

The layers of the vessel vary with the vessel's function. Arteries are thick-walled vessels with large amounts of elastic fibres. The elasticity of these vessels allows them to stretch during cardiac systole and recoil during diastole. The arterioles which are predominantly smooth muscles, serve as resistance vessels for the circulatory system. They constrict or relax as needed to maintain blood pressure.

3.5.2 *Veins and Venules*

The veins are thin-walled, distensible, and collapsible vessels. The structure of the veins allows these vessels to act as a reservoir, or blood storage system.

3.5.3 *Capillaries*

The capillaries are microscopic single-thickness vessels that connect the arterial and venous segments of the circulatory system. Exchange of gases, nutrients, and waste materials occurs through the thin permeable walls of the capillaries.

The approximate dimensions of these human blood vessels can be found in Table 3.3.

Table 3.3: Approximate dimensions of human blood vessels.

Vessel	Diameter (cm)	Length (cm)	Wall Thickness (cm)	Average Velocity (cm/s)
Capillaries	0.0008	0.1	0.0001	0.1
Venules	0.002	0.2	0.0002	0.2
Arterioles	0.005	1	0.02	5
Arteries	0.4	50	0.1	45
Veins	0.5	2.5	0.05	1.0
Aorta	2.5	50	0.2	48
Vena Cava	3.0	50	0.15	38

3.6 Diseases Related to Obstruction of Blood Flow

There are various mechanisms of vessel obstruction. These include (i) thrombus formation, (ii) emboli, (iii) compression, (iv) vasospasm, and (v) structural defects in the vessel (see Fig. 3.7).

3.6.1 *Thrombus Formation*

A thrombus is a blood clot. It can develop in either the arterial or the venous system and obstruct flow. It may cause acute arterial occlusion and, in extremity, will result in sudden onset of acute pain with numbness, tingling, weakness, pallor, and coldness. These changes are rapidly followed by cyanosis, mottling, and loss of sensory, reflex, and motor function. Thromboangiitis obliterans is an inflammatory arterial disorder which causes thrombus formation that affects medium-sized arteries.

Fig. 3.7: Conditions that cause disruption of blood flow.

3.6.2 *Embolus*

An embolus is a foreign mass that is transported in the bloodstream. Although it moves freely in the larger blood vessels, it becomes lodged and obstructs flow once it reaches a smaller vessel. An embolus can be a dislodged thrombus or it can consist of air, fat, tumor cells or other materials. Arterial emboli commonly have their origin in the heart itself and can travel to the brain, spleen, kidney or vessels and the lower extremity before they become lodged and obstruct flow. Acute arterial occlusion normally results from an embolus. Its effects are described above. Emboli tend to lodge in bifurcations of major arteries, including the aorta and iliac arteries.

3.6.3 *Compression*

The lumen of a blood vessel can be occluded by external forces. The external pressure of a tourniquet, for example, is intended to compress blood vessels and interrupt blood flow. Similarly, casts and circular dressings predispose to vessel compression, particularly when swelling occurs after these devices have been applied. Tumors may encroach on blood vessels as they grow. Blood vessels may be compressed between bony structures such as the sacrum and the supporting surfaces of a chair or bed.

3.6.4 *Structural Changes*

Structural changes in blood vessels can take many forms. As explained in Sec. 2.3, arteriosclerosis causes rigidity and narrowing of the arteries. One form of arteriosclerosis, Mönckeberg's medial sclerosis affects the media of medium-sized arteries and is characterised by thickening of the arterioles. The most common form of arteriosclerosis — atherosclerosis, affects both the large and medium-sized arteries. Complications of atherosclerosis include heart attack, aneurysms, stroke, and peripheral vascular disease.

3.6.5 *Vasospasms*

Vasospasm may result from locally or neurally mediated reflexes. Exposure to cold causes severe vasoconstriction in many of the superficial blood vessels. In certain disease states, vasospasm from exposure to cold or other stimuli is excessive and may lead to ischemia and tissue injury. Intense vasospasm of the arteries and arterioles in the fingers and toes causes Raynaud's syndrome, a type of peripheral arterial disease.

Chapter 4

MODELS OF BIOFLUID FLOWS

4.1 Flows in Pipes and Ducts

4.1.1 *Introduction*

As mentioned earlier in Chap. 1, the flows of fluid around objects can be divided into two regions: a thin region close to the object where viscosity effects are important (Boundary Layer Flow), and an outer region where viscosity can be neglected (Potential Flow). Although there is no precise transition line between these two regions, boundary layer is usually defined to be that region where the fluid velocity parallel to the surface is less than 99% of the free stream velocity of the potential flow. The thickness of the boundary layer, δ, grows along a surface over which the fluid is flowing (see Fig. 4.1). Within the boundary layer, the flow is laminar, but as the thickness of the layer grows larger, a transition region appears and the flow in the boundary layer may become turbulent. In fact, if the boundary surface is long enough, the laminar-transition-turbulence sequence occurs in all fluid flows irrespective of the nature of the free stream as to whether it is laminar or turbulent. The Reynolds number for the transition from laminar to turbulent flow is approximately 2300. However, for special conditions transition may occur at Reynolds number as high as 40,000.

Fig. 4.1: Boundary layer flow.

4.1.2 *Developing and Fully Developed Flow*

We shall consider here flow in closed conduits. If a conduit has a circular cross-section of constant area, it is called a pipe. Conduits of other cross-sections but of constant area are called ducts.

The flow in a straight pipe or duct is said to be *fully developed* if the shape of the velocity profile is the same at all cross-sections. Here the velocity varies with distance across the pipe but not along it and there is no well-defined boundary layer. The flow is one directional and there is no velocity perpendicular to the axis of the pipe. We will encounter fully developed flow in the next section, while discussing Poiseuille's flow.

One may think that fully developed flow occurs only under special circumstances. However, many pipe flows are fully developed.

Let us consider the flow from a large reservoir into a pipe (see Fig. 4.2). Consider the flow is laminar in the entry region of the pipe. If the junction between the pipe and the reservoir is nicely rounded, the velocity of the fluid entering the pipe at cross-section 1 will be nearly uniform. However, fluid in contact with the pipe wall will have zero velocity due to the no-slip condition. Hence, a region of slower-moving fluid near the wall is established. This is the boundary layer region which grows wider as the flow moves downstream. Obviously, the boundary layer grows with distance from the entrance until the flow becomes fully developed. The flow in the central region of the pipe is called the *core flow* or the *plug flow* and the region is called the *core region*. Since the fluid near the wall is moving more slowly, the core flow must accelerate as it moves downstream to maintain the constant mass flow at all cross-sections.

It is to be noted that the distinction between laminar and turbulent flow in a pipe or duct applies only to the fully developed flow condition. If the fully developed pipe flow is laminar ($R_e < 2300$), the boundary layer will be laminar, but if the fully developed pipe flow is turbulent ($R_e \geq 4000$, say) then the boundary layer will

Fig. 4.2: Flow in the entry region for laminar flow.

be laminar near the pipe entrance, will undergo transition, and will be turbulent as it approaches the fully developed condition (see Fig. 4.3). For Reynolds number between 2300 and about 4000, the flow is unpredictable, sometimes pulsating or switching back and forth between laminar and turbulent. This type of flow is called *transitional flow*. For $R_e > 4000$, the flow is usually turbulent.

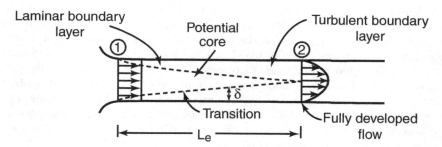

Fig. 4.3: Flow in the entry region for turbulent flow.

Clearly, the boundary layer continues to grow till it fills the entire pipe when the boundary layer thickness is equal to the tube radius. This occurs at section 2 in Figs. 4.2 and 4.3. Downstream of section 2, the flow is fully developed. The flow between sections 1 and 2 is called *developing flow* or *inlet flow* and the region is called the *entry region*. The length L_e till the flow is fully developed is called the *development length* or the *entry length*. This length for laminar flow (Fig. 4.2) as well as for turbulent flow (Fig. 4.3) for a pipe of diameter D was found by analytical and experimental investigations to be

$$L_e = 0.06 R_e D, \quad \text{for laminar flow}, \tag{4.1}$$

$$L_e = 4.4(R_e)^{\frac{1}{6}} D, \quad \text{for turbulent flow}. \tag{4.2}$$

For a laminar flow with Reynolds number $R_e = 2300$, the development length becomes

$$L_e \simeq 140D, \qquad (4.3)$$

whereas for a turbulent flow with Reynolds number $R_e = 10^4$ say, the development length becomes

$$L_e \simeq 20D. \qquad (4.4)$$

For pipes used in many engineering applications which are hundreds or thousands of diameters long, the flow is fully developed over most of their length.

As mentioned earlier, high Reynolds number laminar flows are more common in biological situations than in engineering situations. For example, the flow of air in the respiratory system is usually laminar, but it may become turbulent during heavy breathing or as a result of obstruction in the breathing passageway when the Reynolds number may be as high as 50,000.

The discussion in this section has been confined on flow entering a pipe from a large reservoir. Any disturbance such as placing a valve in the pipe, or changing the pipe diameter will cause the flow to depart from fully developed conditions; however, the flow returns to fully developed conditions after a while downstream.

It is to be noted that in a fully developed flow, because the shape of the velocity profile is constant, the streamlines are straight and parallel to the pipe axis, and the shear stress at the pipe wall is the same in all directions. However, it is to be remembered that the problems of blood flow are much more complicated than the problems of fluid flow in engineering situations because of unusually high Reynolds number of flow. The flow remains laminar at Reynolds numbers as high as 5000-10,000 causing the entry length (which is proportional to the Reynolds number) to be very large so that in most cases the fully developed flow is never reached because tube branchings start before this stage is attained. We will study these problems with much simplification in the following chapters.

4.2 Models of Blood Flows

4.2.1 *Introduction*

As mentioned earlier, interruption of blood flow in either the arterial or venous system interferes with the delivery of oxygen and nutrients to the tissues. Further, it is realised that unfortunately there are numerous arterial diseases which result in the occlusion of blood flow. An appropriate model will assist in the design of a more accurate method to detect these diseases. Thus, the study of models for blood flow has interested clinicians, physiologists, hemorheologists and hydrodynamicists, etc.

The present section mainly aims at finding suitable and more physically and physiologically correct models for the flow of blood. Such models may be helpful in the design of flowmeters which in turn could serve as helpful indicators where detection of diseases is concerned. Alternatively, a knowledge of plausible models can be useful when there is a need to explain perhaps some peculiar clinical or experimental results. There are existing models for blood flows, but they usually contain too many assumptions and hence restrictions on the physical properties of blood. This chapter also examines and explains the inadequacies of such models.

4.2.2 *Poiseuille's Flow*

Having justified the usefulness for a reasonable model of blood flow, we set out to describe some existing models. In this section, we discuss the Poiseuille's law model.

In 1839, the engineer Hagen [35] studied the water flow through brass tubes. Using tubes of known lengths and radii, he found that the volume flow varied as the pressure to a power of the radius and was inversely proportional to the tube length. Further experiments led him to conclude that the power should theoretically be 4. Hagen was thus the first to suggest the importance of r^4 in the flow through any tube.

Poiseuille (1799-1869) was a French physician who first investigated in a quantitative manner the flow of water through glass tubes. Poiseuille's interest was the flow of blood through the vessels of the circulatory system, but worked with water because of the difficulty at that time of preventing blood from clotting on exposure to air. He provided a more complete analysis considering the flow through a circular cylindrical tube. Using compressed air, he forced a constant water volume through capillary tubes, and discovered quite naturally that the volume of water discharged in unit time Q was proportional to the pressure head P and r^4, and inversely proportional to tube length L. This is Poiseuille's law [81, 82] although it is sometimes referred to as Hagen-Poiseuille's law, and mathematically, in the most primitive form:

$$Q = K\frac{Pr^4}{L} \tag{4.5}$$

where K is a constant of proportionality.

The solution was derived by Wiedemann (1856) and Hagenbach (1860) independently. The derivation is given below:

Consider a tube of length L and radius R (Fig. 4.4) in which a viscous fluid of viscosity μ is flowing. Let p_1 and p_2 be the inflow pressure at A and the outflow pressure at B respectively, and let these be maintained constant throughout our discussion. The flow is assumed to be such that each particle of liquid moves parallel to the axis with a constant velocity. All points lying on the same circle around the axis will hence have the same velocity, so the liquid is thought of as having cylindrical laminae moving at velocities u, u being functions of their radii. This is actually Newton's hypothesis for laminar flow.

The force exerted by the difference in pressure at the ends on a cross-sectional

Fig. 4.4: Flow in a circular cylindrical tube.

area is:

$$F = \pi r^2 (p_1 - p_2). \tag{4.6}$$

The retarding force to this motion is the viscous force F_v given by:

$$\text{area} \times \text{viscosity} \times \text{velocity gradient}.$$

Hence

$$F_v = (2\pi r L) \times \mu \times \frac{du}{dr}. \tag{4.7}$$

These forces are equal and opposite, hence,

$$F = -F_v$$

i.e., $\pi r^2 (p_1 - p_2) = -2\pi r L \mu \frac{du}{dr}$

or $\quad r(p_1 - p_2) = -2L\mu \frac{du}{dr}$

i.e., $\quad \dfrac{du}{dr} = \dfrac{-r(p_1 - p_2)}{2L\mu}. \tag{4.8}$

Equation (4.8) is thus the expression for the velocity gradient, and on integration, yields

$$u = \frac{-r^2(p_1 - p_2)}{4L\mu} + C, \tag{4.9}$$

where C is an arbitrary constant.

If we assume that the fluid in contact with the tube wall is at rest, then $u = 0$ at $r = R$, hence,

$$C = \frac{R^2(p_1 - p_2)}{4L\mu}. \tag{4.10}$$

Substituting Eq. (4.10) into Eq. (4.9), we get,

$$u = \frac{p_1 - p_2}{4L\mu}(R^2 - r^2). \tag{4.11}$$

Equation (4.11) is an equation for a parabola (see Fig. 4.5). Hence, the volume of the paraboloid formed by revolving this parabola around its axis is the volume of fluid Q which flows in unit time, i.e.,

$$Q = \int_0^R 2\pi u.r \, dr. \tag{4.12}$$

Substitution of Eq. (4.11) into Eq. (4.12) gives

$$Q = \frac{2\pi(p_1 - p_2)}{4L\mu} \int_0^R (R^2 - r^2)r \, dr;$$

$$\text{or} \quad Q = \frac{(p_1 - p_2)\pi R^4}{8L\mu}. \tag{4.13}$$

Equation (4.13) is commonly known as Poiseuille's equation, or Poiseuille's law.

Fig. 4.5: Velocity profile using Poiseuille's law.

In the above derivation of Poiseuille's equation, which gives a plausible expression for blood flow, several conditions have to be satisfied. Amongst these conditions, the two most important are:

(1) The tube is rigid; its diameter remains constant with respect to internal pressures p_1 and p_2. We know from our previous discussion that blood vessels, in particular, arteries, are not rigid but elastic, and they distend with increasing internal pressure. This distensibility is much less in smaller vessels, in which case Poiseuille's equation may apply.

(2) The fluid is homogeneous; i.e., having a constant viscosity at all rates of shear (velocity gradients). Such fluids are Newtonian fluids. We have seen that blood is not homogeneous but a suspension of cells in plasma. Hence, blood does not have a constant viscosity, and this fact is ignored by Eq. (4.13) which includes the constant μ.

At high flow rates, cells are observed to move away from the walls of the blood vessels towards the axis, leaving a plasma layer which leads to deviations from Poiseuille's equation. At low flow rates, the cells aggregate and move as clumps, so the deviations are not as pronounced. In the case where there is no marginal layer and yield stress effects are negligible, this equation is exactly correct. This is also true when the tube diameter is less than 0.05 cm.

Although it is now recognised that blood is non-Newtonian in nature, the works of Hagen and Poiseuille are not forgotten and are still used for their simplicity.

4.3 Consequence of Poiseuille's Flow

Poiseuille's law is so well-established experimentally, that it is often used in order to determine the viscosity coefficient μ of viscous fluids. When blood was examined in this manner, it was found that the viscosity of blood $\mu_B = 5\mu_0$ (viscosity of water), if the diameter of the tube is relatively large. Thus at normal physiological temperature of 37°C, $\mu_0 = 0.07P\,(P = \text{Poise} = \text{dyne sec cm}^{-2})$ and $\mu_B = 0.035P$, as determined by Poiseuille's law in large tubes.

The fact that the effective viscosity coefficient of blood according to Poiseuille's law depends on the radius of the tube in which it is measured indicates further that blood is not a Newtonian fluid, for which μ is constant.

4.4 Applications of Poiseuille's Law for the Study of Blood Flow

We have seen from Poiseuille's law,

$$Q = \frac{\pi R^4}{8\mu L}(p_1 - p_2),$$

that the flux Q is proportional to the pressure difference $(p_1 - p_2)$. However, the observed flow of blood through the veins shows a marked non-linearity (see Fig.

4.6). For small pressure differences, the flux Q is proportional to $(p_1 - p_2)$ just as in Poiseuille's law, but when the pressure difference becomes large, the flux gradually attains a maximum value and then no longer increases.

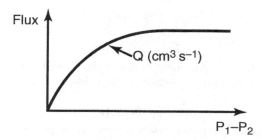

Fig. 4.6: Relationship between flux and pressure-difference in Poiseuille's flow.

A similar situation occurs in the flow through a rubber tube. This result suggests that it is important to take into account the elastic nature of the blood carrying vessels.

An artery or a vein is not a rigid tube but is rather an elastic one. Consequently, the radius of such a tube is variable. Imagine such a vessel filled with fluid at rest and surrounded by fluid (see Fig. 4.7). Let the wall thickness of the tube be h, the radius of the tube r, the exterior pressure p_0, and the interior fluid pressure be p. The change in p, namely, $(p - p_0)$, is called the *transmural pressure difference*. Under normal circumstances the thickness h is small compared to the resting radius (when exterior pressure p_0 equals the interior pressure p) of the vessel. Consequently, as a good approximation we can treat the wall as a thin elastic membrane.

Fig. 4.7: An elastic tube filled with fluid at rest and surrounded by fluid.

Let T denote the tension per unit length of the tube and per unit thickness. Consider the equilibrium of half of such a long cylinder vessel plus the fluid contained within it. The net downward force per unit length on this half cylinder is $2Th$ and it is balanced by the net upward force per unit length which is given by

$$\int_0^\pi (p - p_0) r \sin \theta \, d\theta, \tag{4.14}$$

which equals

$$2r(p - p_0). \tag{4.15}$$

Thus we have

$$Th = r(p - p_0). \tag{4.16}$$

The tension T that develops is a property of the elastic wall in its reaction to stretch. The force per unit acting on the surface of the volume element is called stress (see Fig. 4.8), i.e.,

$$\mathbf{F} = \mathbf{N} + \mathbf{H}, \tag{4.17}$$

where $\mathbf{N}$ is the normal component such as tension or pressure called the normal stress, and $\mathbf{H}$ is the tangential component called the shearing stress. In response to stress the volume element undergoes a deformation called strain. We have similarly normal strain and shearing strain.

According to Hooke's law, we have

$$T = E\frac{(r - r_0)}{r_0} \tag{4.18}$$

where $r = r_0$ in equilibrium position when the tension T is zero.

Fig. 4.8: Stress components in an elastic membrane.

Let us now consider the steady flow through an elastic tube of length L such as an artery or a vein. A diagram of the structure is given in Fig. 4.9.

We assume that the tube is long and the pressure is a function of x alone and that

$$p(x)|_{x=0} = p_1 \quad \text{and} \quad p(x)|_{x=L} = p_2 \tag{4.19}$$

where p_1 and p_2 are the inlet and outlet pressures respectively. The external pressure of the fluid surrounding the tube is assumed to have a constant value of p_0 (say). We have from Eq. (4.16) that

$$p(x) - p_0 = \frac{Th}{r} \tag{4.20}$$

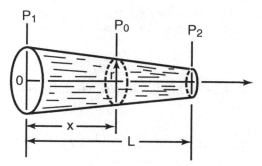

Fig. 4.9: Flow through an elastic tube.

where r is the cross-sectional radius at location x. The pressure-radius relation is obtained from Eqs. (4.18) and (4.20) giving,

$$p(x) - p_0 = \frac{Eh}{r} \frac{(r - r_0)}{r_0} = \frac{Eh}{r_0} \left(1 - \frac{r_0}{r}\right). \qquad (4.21)$$

Now the flow through the tube is assumed to obey Poiseuille's law in the form

$$Q = -\frac{\pi}{8\mu} \frac{dp}{dx} r^4, \qquad (4.22)$$

where

$$\frac{dp}{dx} = \frac{p_2 - p_1}{L}, \qquad (4.23)$$

and r and p are related by Eq. (4.21). From Eq. (4.21) we see that r is a function of $(p - p_0)$, i.e.,

$$r = f(p - p_0) = r(p - p_0)(\text{say}). \qquad (4.24)$$

Integrating Eq. (4.22) we obtain

$$\int_0^L Q \, dx = -\frac{\pi}{8\mu} \int_{p_1}^{p_2} r^4 \, dp, \qquad (4.25)$$

which reduces to

$$QL = \frac{\pi}{8\mu} \int_{p_2}^{p_1} r^4 \, dp. \qquad (4.26)$$

Making the substitution $p' = p - p_0$ and simplifying the resulting equation, gives for the volume flux

$$Q = \frac{\pi}{8\mu L} \int_{p_2 - p_0}^{p_1 - p_0} r^4 \, dp', \qquad (4.27)$$

where r is a function of p'. Hence we see that Q is proportional to $1/L$ and is a function of $(p_1 - p_0)$ and $(p_2 - p_0)$. To evaluate the above integral, Eq. (4.21) is rearranged to give

$$r = \frac{r_0}{(1 - \frac{r_0}{Eh}p')}. \tag{4.28}$$

Therefore, with the use of Eq. (4.28), the definite integral appearing in Eq. (4.27) can easily be evaluated to give

$$Q = \frac{\pi r_0^3 Eh}{24\mu L}\left[\left(1 - \frac{r_0}{Eh}(p_1 - p_0)\right)^{-3} - \left(1 - \frac{r_0}{Eh}(p_2 - p_0)\right)^{-3}\right], \tag{4.29}$$

or

$$Q \approx \frac{\pi r_0^3 Eh}{24\mu L}\left[\left(1 + \frac{3r_0}{Eh}(p_1 - p_0) + \cdots\right) - \left(1 + \frac{3r_0}{Eh}(p_2 - p_0) + \cdots\right)\right] \tag{4.30}$$

which is certainly not linear. By retaining only the first order terms, the above collapses to Poiseuille's law. We know that r^4 is a function of p' (i.e., $r^4(p') = r^4(p - p_0)$). When $(p_1 - p_2)$ is small, r^4 is not too rapidly varying and can be taken as a constant for this period of integration and we have from Eq. (4.27),

$$Q \approx \frac{\pi}{8\mu L}r^4[(p_1 - p_0) - (p_2 - p_0)]$$

$$\approx \frac{\pi}{8\mu L}r^4(p_1 - p_2), \tag{4.31}$$

which like Poiseuille's law is proportional to the pressure difference $(p_1 - p_2)$.

4.5 Pulsatile Flow

Consider the flow in a circular vessel. Assuming the only non-zero component of the velocity vector is in the axial direction and is denoted by u. Using cylindrical polar coordinates, and taking the problem to be axisymmetric (i.e., $u = u(r, x, t), p = p(r, x, t)$), the continuity equation reduces to

$$\frac{\partial u}{\partial x} = 0. \tag{4.32}$$

The Navier-Stokes equations become

$$\rho\frac{\partial u}{\partial t} = -\frac{\partial p}{\partial x} + \frac{\mu}{r}\frac{\partial}{\partial r}\left(r\frac{\partial u}{\partial r}\right), \tag{4.33}$$

$$0 = -\frac{\partial p}{\partial r}. \tag{4.34}$$

The above Eqs. (4.32) and (4.34) imply that

$$u = u(r,t) \quad \text{and} \quad p = p(x,t). \tag{4.35}$$

Therefore Eq. (4.33), governing the flow simplifies to

$$\frac{\mu}{r}\frac{\partial}{\partial r}\left(r\frac{\partial u}{\partial r}\right) - \rho\frac{\partial u}{\partial t} = \frac{\partial p}{\partial x}. \tag{4.36}$$

Observing that the left-hand side of the above equation is a function of (r,t) only and the right-hand side a function of (x,t) only, we conclude that both sides must be a function of t only.

Consider now a pulsatile sinusoidal flow with the pressure gradient and the axial velocity as

$$\frac{\partial p}{\partial x} = -Pe^{i\omega t}, \tag{4.37}$$

$$u(r,t) = U(r)e^{i\omega t} \tag{4.38}$$

where P is a constant, and $U(r)$ is the velocity profile across the tube of radius a. We assume that the flow is identical at each section along the tube so that a travelling wave solution can be neglected. It is clear that when $\omega = 0$ (the steady case), the flow becomes that of Poiseuille flow discussed earlier.

From Eqs. (4.37) and (4.38) it is clear that the real part gives the velocity for the pressure gradient $P\cos\omega t$ and the imaginary part gives the velocity for the pressure gradient $P\sin\omega t$. Upon substituting Eqs. (4.37) and (4.38) into Eq. (4.36) the following equation results:

$$\frac{d^2U}{dr^2} + \frac{1}{r}\frac{dU}{dr} - \frac{i\omega\rho}{\mu}U = -\frac{P}{\mu}. \tag{4.39}$$

It is known that the general solution of an ordinary differential equation in the form

$$\frac{d^2y}{dx^2} + \frac{1}{x}\frac{dy}{dx} - K^2y = 0 \tag{4.40}$$

is

$$y = AJ_0(iKx) + BY_0(iKx), \tag{4.41}$$

which involves Bessel functions of complex argument. Thus, the solution of Eq. (4.39) is

$$U(r) = AJ_0\left(i\sqrt{(i\omega\rho/\mu)}r\right) + BY_0\left(i\sqrt{(i\omega\rho/\mu)}r\right) + \frac{P}{\omega\rho i}. \tag{4.42}$$

As U must be finite on the axis (i.e., at $r = 0$), and since $Y_0(0)$ is not finite, then B has to be zero. Also, because of the no-slip condition $U(r)|_{r=a} = 0$, we have

$$AJ_0\left(i^{3/2}\sqrt{(\omega\rho/\mu)}a\right) + \frac{P}{\omega\rho i} = 0. \tag{4.43}$$

Let us now introduce a non-dimensional parameter α, known as the *Womersley parameter* given by

$$\alpha = a\sqrt{\frac{\omega\rho}{\mu}} \quad \text{or} \quad \alpha = a\sqrt{\frac{\omega}{\nu}} \tag{4.44}$$

where $\nu = \mu/\rho$ is the kinematic viscosity. From Eq. (4.43)

$$A = -\frac{P}{\omega\rho i}\frac{1}{J_0(i^{3/2}\alpha)} \quad \text{or} \quad A = \frac{iP}{\omega\rho}\frac{1}{J_0(i^{3/2}\alpha)}, \tag{4.45}$$

and finally, from Eq. (4.42) we get

$$U(r) = -\frac{iP}{\omega\rho}\left(1 - \frac{J_0(i^{3/2}\alpha r/a)}{J_0(i^{3/2}\alpha)}\right). \tag{4.46}$$

In the limit as α approaches zero, i.e., as $\omega \to 0$, the velocity profile becomes parabolic. As α tends to infinity, i.e., viscosity is unimportant ($\mu \to 0$), it can be shown that

$$\frac{J_0(i^{3/2}\alpha r/a)}{J_0(i^{3/2}\alpha)} \to 0 \tag{4.47}$$

which implies that

$$U(r) \to -\frac{iP}{\omega\rho}. \tag{4.48}$$

In fact, introducing the idea of a *Stokes boundary layer*, of thickness δ proportional to $1/\alpha$, it can be concluded that in this case the boundary layer thickness at the cylinder wall disappears.

It is interesting to note that the above expression is independent of viscosity, μ, and exactly $90°$ out of phase with P. Actually this result corresponds to the Euler equation (which only holds for $\mu = 0$, i.e., inviscid fluids) namely,

$$\frac{\partial u}{\partial t} = -\frac{1}{\rho}\frac{\partial p}{\partial x}. \tag{4.49}$$

Using Eqs. (4.37) and (4.38) in Eq. (4.49) we easily see that

$$U = \frac{P}{i\omega\rho} = -\frac{iP}{\omega\rho} \tag{4.50}$$

which coincides with the result in Eq. (4.48).

Hence, the final result for the velocity of pulsatile flow in a cylindrical tube of radius a is

$$u(r,t) = -\frac{iP}{\omega\rho}\left(1 - \frac{J_0(i^{3/2}\alpha r/a)}{J_0(i^{3/2}\alpha)}\right)e^{i\omega t}. \tag{4.51}$$

The volumetric flow rate Q, is given by

$$Q = \int_0^a u 2\pi r \, dr$$

$$= -\frac{2\pi i P}{\omega \rho} e^{i\omega t} \left[\int_0^a r \, dr - \frac{1}{J_0(i^{3/2}\alpha)} \int_0^a r J_0\left(i^{3/2}\alpha r/a\right) dr \right]$$

$$= -\frac{\pi i P}{\omega \rho} e^{i\omega t} a^2 \left[1 - \frac{2}{a^2 J_0(i^{3/2}\alpha)} \int_0^a r J_0\left(i^{3/2}\alpha r/a\right) dr \right]. \qquad (4.52)$$

Now consider the term

$$\frac{2}{a^2 J_0(i^{3/2}\alpha)} \int_0^a r J_0\left(i^{3/2}\alpha r/a\right) dr, \qquad (4.53)$$

by making the substitutions

$$i^{3/2}\alpha = \beta \quad \text{and} \quad r\beta/a = \theta \qquad (4.54)$$

the above term simplifies to

$$\frac{2}{\beta^2 J_0(\beta)} \int_0^\beta \theta J_0(\theta) \, d\theta, \qquad (4.55)$$

but

$$\int_0^\beta \theta J_0(\theta) \, d\theta = \beta J_1(\beta). \qquad (4.56)$$

Therefore Eq. (4.55) reduces to

$$\frac{2 J_1(\beta)}{\beta J_0(\beta)} \qquad (4.57)$$

and Eq. (4.52) can be written as

$$Q = -\frac{\pi i P e^{i\omega t}}{\omega \rho} a^2 \chi(\beta), \qquad (4.58)$$

where

$$\chi(\beta) = 1 - \frac{2 J_1(\beta)}{\beta J_0(\beta)}. \qquad (4.59)$$

For small values of β (and hence α) the above expression can be expanded in a Taylor series to give,

$$\chi(\beta) = 1 - \frac{2}{\beta} \frac{\left(\frac{\beta}{2} - \left(\frac{\beta}{2}\right)^3 / 1!2! + \left(\frac{\beta}{2}\right)^5 / 2!3! - \cdots \right)}{\left(1 - \left(\frac{\beta}{2}\right)^2 / (1!)^2 + \left(\frac{\beta}{2}\right)^4 / (2!)^2 - \cdots \right)}$$

$$= 1 - \frac{1 - \beta^2/8 + \cdots}{1 - \beta^2/4 + \cdots}$$

$$= 1 - (1 - \beta^2/8 + \cdots)(1 + \beta^2/4 - \cdots) = -\beta^2/8 + O(\beta^4). \qquad (4.60)$$

By substituting $\beta = i^{3/2}\alpha$ into the above equation we get

$$\chi(\beta) = \chi(\alpha) = \frac{i\alpha^2}{8} + O(\alpha^4). \qquad (4.61)$$

Then substituting this result into Eq. (4.58) and using the definition of α as given in Eq. (4.44), we finally obtain an expression for Q valid for small α, namely

$$Q = \left(\frac{P\pi a^4}{8\mu} + O(\alpha^4)\right) e^{i\omega t}. \qquad (4.62)$$

Again as $\alpha \to 0$ then $\omega \to 0$ and $Q \to Q_0$ where

$$Q_0 = \frac{\pi P a^4}{8\mu} e^{i\omega t} \qquad (4.63)$$

and

$$|Q_0| = \frac{\pi P a^4}{8\mu} \qquad (4.64)$$

where $|Q_0|$ is the volumetric flow rate for a constant pressure gradient and is the same as for steady Poiseuille's flow.

By separating Eq. (4.58) into its real and imaginary parts we get

$$Q = \frac{\pi P a^2}{\omega\rho} \{[\chi_2(\alpha)\cos\omega t + \chi_1(\alpha)\sin\omega t] - i[\chi_1(\alpha)\cos\omega t - \chi_2(\alpha)\sin\omega t]\}, \quad (4.65)$$

where

$$\chi(\alpha) = \chi_1(\alpha) + i\chi_2(\alpha). \qquad (4.66)$$

The real part of Eq. (4.65) gives the flux when the pressure is $P\cos\omega t$ and the imaginary part gives the flux when it is $P\sin\omega t$.

4.6 Further Discussion on Pulsatile Flow

We are interested in two quantitative aspects of pulsatile flow namely,

(i) The magnitude of Q,

$$|Q| = \frac{\pi P a^2}{\omega\rho}\sqrt{\chi_1^2(\alpha) + \chi_2^2(\alpha)} = \frac{\pi P a^4}{\mu\alpha^2}\sqrt{\chi_1^2(\alpha) + \chi_2^2(\alpha)}. \qquad (4.67)$$

(ii) The phase angle between the pressure gradient $Pe^{i\omega t}$ and the flow rate Q, given by (from Eq. (4.58)),

$$\tan\phi = \frac{\chi_1}{\chi_2}.\tag{4.68}$$

From Eq. (4.59),

$$\chi_1 + i\chi_2 = 1 - \frac{2}{i^{3/2}\alpha}\frac{J_1(i^{3/2}\alpha)}{J_0(i^{3/2}\alpha)}.\tag{4.69}$$

The graphs of the values of χ_1 and χ_2 from a table of Bessel functions are given in Fig. 4.10.

Fig. 4.10: Graphs of $\chi_1(\alpha)$ and $\chi_2(\alpha)$.

We also have

$$\frac{|Q|}{|Q_0|} = \frac{8\sqrt{\chi_1^2 + \chi_2^2}}{\alpha^2}.\tag{4.70}$$

Since the heart beats about 72 times per minute, we can take

$$\frac{2\pi}{\omega} = \frac{60}{72} \quad \text{or} \quad \omega \approx 8\,\text{rad sec}^{-1}.$$

For blood, taking $\rho = 1.05$ gm cc^{-1}, $\mu = 4$ cp $= 0.04$ gm cm^{-1}sec^{-1}, and $a = 0.5$ cm, we get $\alpha \approx 7$. Even for this value, the volumetric flow rate $|Q|$ is about one-eighth of the steady state value. Relationships of α with the flow rate for various artery sizes are shown in Table 4.1.

Table 4.1: Relationships of artery size, flow rate and Womersley parameters are shown for the pressure gradient and blood viscosity being taken as 100 gm cm^{-1} sec^{-2} and 0.04 gm cm^{-1}sec^{-1} respectively.

Womersley Parameter, λ	Artery Radius, a (cm)	Flow Rate, $\|Q\|$ (cm^{3}sec^{-1})
2	0.100	0.083
2	0.200	1.336
2	0.400	21.373
2	0.500	52.180
4	0.100	0.036
4	0.200	0.578
4	0.400	9.251
4	0.500	22.587
6	0.100	0.017
6	0.200	0.276
6	0.400	4.417
6	0.500	10.784
8	0.100	0.010
8	0.200	0.165
8	0.400	2.633
8	0.500	5.429
10	0.100	0.007
10	0.200	0.109
10	0.400	1.737
10	0.500	4.242

Introducing an impedance parameter z defined by

$$z = \frac{-\frac{\partial p}{\partial x}}{Q} = \frac{Pe^{i\omega t}}{Q}, \tag{4.71}$$

which approaches z_0 as $\alpha \to 0$ (i.e., $\omega \to 0$) where $z_0 = \frac{P}{Q_0}$. We thus have

$$\frac{|z|}{z_0} = \frac{Q_0}{|Q|}. \tag{4.72}$$

The impedance parameter z introduced in Eq. (4.71) is usually called the longitudinal impedance parameter. The idea of impedance is analogous to the concept of transfer function in a mechanical and electrical system. In fact, if the ratio between the pressure gradient and the flow is used to calculate impedance z as defined by Eq. (4.71), it must then be characterised by both a magnitude and an angle, and thus can be written as $z = |z|e^{i\theta}$.

This phase difference, θ, is very important, since it determines the work done by the system, obtained by the product of pressure and rate of flow. The impedance is a function of position, but not a function of time. If it is used to determine the apparent phase velocity then the speed of propagation will also be a function of position. However, it is highly complex to calculate the impedance as it involves a measurement of both pressure and flow. In practice, the pressure gradient-flow relation has usually been taken from the theoretical equations and used in the reverse direction to calculate pulsatile flow (see Fry [27]).

4.7 The Pulse Wave

It is a well-established fact that the flow of blood is pulsatile as a consequence of the beating of the heart. This beating produces a pressure wave that travels through the blood and this pressure wave is the pulse felt in the wrist. In practice, the heart beats about a constant period T of approximately 1 second. We are interested here to obtain the velocity of this pulse wave.

Consider an elastic tube through which the fluid is flowing. Assume now that the fluid is both inviscid and incompressible. Let $A(x,t)$ be the cross-sectional area at a distance x and at any time t, $u(x,t)$ the velocity parallel to the tube, and ρ the density of the fluid. The net force in the positive x direction on the volume element (see Fig. 4.11) is

$$pA - \left[pA + \frac{\partial}{\partial x}(pA)\Delta x\right] + p_0 \frac{\partial A}{\partial x}\Delta x = -\frac{\partial}{\partial x}[(p - p_0)A]\Delta x. \qquad (4.73)$$

Now this must be equal to the mass times acceleration of the volume element given by,

$$\rho A \Delta x \left(\frac{\partial u}{\partial t} + u\frac{\partial u}{\partial x}\right). \qquad (4.74)$$

Thus the equation of motion for the fluid through the pipe is

$$\rho A \left(\frac{\partial u}{\partial t} + u\frac{\partial u}{\partial x}\right) = -\frac{\partial}{\partial x}\left((p - p_0)A\right). \qquad (4.75)$$

The continuity equation for an incompressible fluid is

$$\frac{\partial A}{\partial t} + \frac{\partial}{\partial x}(Au) = 0, \qquad (4.76)$$

and the pressure-radius elastic relation is

$$p - p_0 = \frac{Eh}{r_0}\left(1 - \frac{r_0}{r}\right) = \frac{Eh}{r_0}\left(1 - \left(\frac{A_0}{A}\right)^{1/2}\right), \qquad (4.77)$$

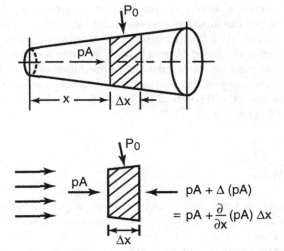

Fig. 4.11: Pressure wave and volume element of fluid flowing in the x-direction.

where $A = \pi r^2$ and $A_0 = \pi r_0^2$.

We linearise Eqs. (4.75)–(4.77) by assuming that u, $(p - p_0)$ and $(A - A_0)$ (i.e., $A = A_0 + h$, where h is some small parameter) together with their derivatives are small. This approach leads to the following three equations:

$$\rho \frac{\partial u}{\partial t} = -\frac{\partial p}{\partial x}, \tag{4.78}$$

$$\frac{\partial A}{\partial t} + A_0 \frac{\partial u}{\partial x} = 0, \tag{4.79}$$

$$p - p_0 = \frac{Eh}{2r_0 A_0}(A - A_0). \tag{4.80}$$

Differentiating Eq. (4.78) with respect to x and Eq. (4.79) with respect to t, we can eliminate u from both equations to give

$$\frac{\partial^2 A}{\partial t^2} = \frac{A_0}{\rho} \frac{\partial^2 p}{\partial x^2}. \tag{4.81}$$

Now from Eq. (4.80) we obtain

$$\frac{\partial^2 p}{\partial t^2} = \frac{Eh}{2r_0 A_0} \frac{\partial^2 A}{\partial t^2}. \tag{4.82}$$

Therefore on combining these two equations we finally obtain

$$\frac{\partial^2 p}{\partial x^2} = \frac{1}{c^2} \frac{\partial^2 p}{\partial t^2} \tag{4.83}$$

where $c^2 = Eh/2\rho r_0$. Equation (4.83) is the classical wave equation with wave speed c. It can be seen that $u(x, t)$ also satisfies the wave equation given by,

$$\frac{\partial^2 u}{\partial x^2} = \frac{1}{c^2}\frac{\partial^2 u}{\partial t^2}.$$ (4.84)

We will discuss this equation in the next section.

4.8 Mones-Korteweg Expression for Wave Velocity in an Inviscid Fluid-Filled Elastic Cylindrical Tube

We need to employ the governing fluid-dynamical equations, namely the continuity and momentum equations, the governing equations of tube motion and the matching kinematic boundary conditions. The following assumptions are incorporated in the basic equations:

(i)   Viscosity is neglected.

(ii) The axial flow velocity is small relative to the pulse wave velocity (of the order of less than 10 per cent).

(iii) The vessel diameter is likewise an order of magnitude smaller than the wavelength.

Let u and v denote the flow velocities in the axial and radial directions (x and r) respectively. By examining each equation separately, the following results are obtained:

(a) *Continuity Equation*

The continuity equation is given by,

$$\frac{\partial u}{\partial x} + \frac{1}{r}\frac{\partial(rv)}{\partial r} = 0.$$ (4.85)

We now introduce the average axial flow velocity

$$\bar{u}(x, t) = \frac{1}{\pi a^2}\int_0^a u(x, r, t)2\pi r\, dr,$$ (4.86)

where a is the tube radius.

Upon multiplying both terms of the continuity equation by $2\pi r dr$, integrating with respect to r, and imposing the following boundary conditions:

$$v(x, r, t)|_{r=0} = 0, \quad \text{(due to axisymmetrical flow)}$$ (4.87a)

and
$$v(x,r,t)|_{r=a} = v_w, \quad \text{(wall velocity)} \tag{4.87b}$$

we finally obtain,
$$v_w(x,t) = -\frac{a}{2}\frac{\partial \bar{u}(x,t)}{\partial x}. \tag{4.88}$$

(b) *Momentum Equations*

The momentum equations are,

$$\rho\left(\frac{\partial u}{\partial t} + u\frac{\partial u}{\partial x} + v\frac{\partial u}{\partial r}\right) = -\frac{\partial p}{\partial x}, \tag{4.89a}$$

$$\rho\left(\frac{\partial v}{\partial t} + u\frac{\partial v}{\partial x} + v\frac{\partial v}{\partial r}\right) = -\frac{\partial p}{\partial r}. \tag{4.89b}$$

As a result of our assumptions, the radial flow velocity and the convective acceleration terms are respectively of smaller order of magnitude relative to the axial flow velocity and the local acceleration terms. The axial and radial momentum equations now result in:

$$\rho\frac{\partial \bar{u}(x,t)}{\partial t} = -\frac{\partial p(x,t)}{\partial x}, \tag{4.90}$$

$$\frac{\partial p}{\partial r} = 0, \tag{4.91}$$

where ρ is the fluid density and the average axial velocity $\bar{u}$ is as defined in Eq. (4.86).

(c) *Tube Motion Equation*

By considering the dynamic equilibrium of a cylindrical tube element in the radial direction, and by neglecting the inertial force to the motion of the tube wall element (of circumferential and axial dimensions $ad\theta$ and dx respectively, and wall thickness h), we obtain
$$pad\theta dx - (\sigma h dx)d\theta = 0. \tag{4.92}$$

By putting
$$\sigma = Ee = E(\eta/a), \tag{4.93}$$

where η is the wall displacement, and E the wall material elastic modulus, and simplifying Eqs. (4.92) and (4.93), we get

$$\eta = \frac{a^2}{hE}p(x,t). \tag{4.94}$$

(d) *Kinematic Matching Boundary Conditions*

By requiring that the fluid velocity at the wall, v_w, equals the wall motion velocity ($\dot{\eta}$), and assuming the variation of wall displacement η in the axial direction is negligible we obtain

$$v_w(x, t) = \dot{\eta}(x, t). \tag{4.95}$$

By combining Eqs. (4.88), (4.90), (4.94) and (4.95), we obtain the pulse wave equations, namely,

$$\frac{\partial^2 \bar{u}}{\partial x^2} = \frac{1}{c^2} \frac{\partial^2 \bar{u}}{\partial t^2} \tag{4.96a}$$

and

$$\frac{\partial^2 p}{\partial x^2} = \frac{1}{c^2} \frac{\partial^2 p}{\partial t^2}, \tag{4.96b}$$

where the pulse wave velocity, c, is given by

$$c = \left(\frac{Eh}{2a\rho} \right)^{1/2}. \tag{4.97}$$

4.9 Applications in the Cardiovascular System

The above analysis with the derived expression for the pulse wave velocity can provide the basis for characterising the modulus of elasticity (Young's modulus) as a function of its inner radius, as well as for demonstrating the construction of the pressure waveform from the flow-velocity waveform and vice versa.

The expression (4.97) for the pulse wave velocity enables the determination of the arterial modulus E (as a function of a) by ultrasound measurement (sector scan echocardiography) of the values of arterial dimensions (a, h), the waveforms of the arterial inner radius at two sites, the transit time (as the time interval between the waveform peaks or centroids), and hence the pulse wave velocity. The present day resolution of sector scan echocardiography systems can realistically enable us to determine only the aortic pressure with reasonable accuracy.

The solution to Eq. (4.96a) can be written, in d'Alembert's form as,

$$\bar{u} = f(x - ct) + g(x + ct). \tag{4.98}$$

The functions f and g represent waves travelling in the positive and negative x directions respectively with speed c.

Neglecting the reflected wave, and using Eq. (4.90), we get

$$\bar{u} = f(x - ct) \tag{4.99a}$$

and

$$p = \rho c f(x - ct). \tag{4.99b}$$

Thus, the arterial pressure and velocity are proportional to each other; this explains why the pressure and velocity have similar waveforms. This relation can provide the basis of noninvasive (without any surgical intervention) measurement of arterial pressure waveform from noninvasive pulsed Doppler flowmeter measurement of the averaged blood flow velocity, $\bar{u}$, waveform; and the noninvasive measurement of the pulse wave velocity as mentioned above.

If the $\bar{u}(x, t)$ waveform is measurable by means of the pulsed Doppler flowmeter, then it can be expressed in Fourier series form by computing the Fourier coefficients, u_n, by means of the Euler integral formula; this is illustrated below:

Assume

$$\bar{u}(x, t) = \sum_{n=-\infty}^{\infty} u_n(x) e^{(i2\pi nt)/T} \tag{4.100}$$

such that

$$u_n = \frac{1}{T} \int_0^T \bar{u}(x, t) e^{-(i2\pi nt)/T} \, dt, \tag{4.101}$$

where T is the period of the heart beat or cardiac cycle.

Let the corresponding pressure waveform be given by

$$p(x, t) = \sum_{n=-\infty}^{\infty} p_n(x) e^{(i2\pi nt)/T} \tag{4.102}$$

where

$$p_n = \frac{1}{T} \int_0^T p(x, t) e^{-(i2\pi nt)/T} \, dt. \tag{4.103}$$

Since from Eq. (4.99)

$$p_n = \rho c u_n, \tag{4.104}$$

we can construct the pressure waveform $p(x, t)$ corresponding to the pulsed Doppler flowmetry derived $\bar{u}(x, t)$ waveform, by following the steps given below:

(i) Determine p_n, associated with u_n of $\bar{u}(x, t)$ corresponding to each constituent frequency parameter $\omega_n = 2\pi n/T$.

(ii) Obtain the composite waveform $p(x, t)$ according to Eq. (4.102).

(iii) Calibrate the waveform by having the peak waveform amplitude equal to the systolic pressure measured by sphygmomanometry.

Thus, this analysis enables noninvasive and continuous determination of the pressure waveform. By studying the hitherto unavailable aortic pressure waveforms of different diseases of the cardiovascular system, we can associate waveform parameters with cardiovascular abnormalities and develop a new cardiovascular diagnostic field, based on the pressure waveform analysis.

4.10 Wave Propagation Accounting for Viscosity and its Application to Cardiac Output Determination

This theory incorporates the effect of viscosity, μ, in the analysis of pulse wave propagation in viscous fluid-filled cylindrical tubes. The other assumptions stated in Sec. 4.8 are, however, still relevant.

The equations of continuity and momentum for a viscous fluid are as follows:

$$\frac{\partial u}{\partial x} + \frac{1}{r}\frac{\partial(rv)}{\partial r} = 0, \tag{4.105}$$

$$\rho\frac{\partial u}{\partial t} + \frac{\partial p}{\partial x} = \mu\left(\frac{\partial^2 u}{\partial r^2} + \frac{1}{r}\frac{\partial u}{\partial r}\right), \tag{4.106}$$

$$\frac{\partial p}{\partial r} = 0, \tag{4.107}$$

where u and v are the axial and radial (x and r) flow velocities repectively.

The elastic tube motion equations are (see Fung [28]):

$$\rho_w h\frac{\partial^2 \eta}{\partial t^2} = \sigma_{rr} - \frac{Eh}{1-\nu^2}\left(\frac{\eta}{a^2} + \frac{\nu}{a}\frac{\partial\xi}{\partial x}\right); \quad \sigma_{rr} = p, \tag{4.108}$$

$$\rho_w h\frac{\partial^2 \xi}{\partial t^2} = -\sigma_{rx} - \frac{Eh}{1-\nu^2}\left(\frac{\partial^2\xi}{\partial x^2} + \frac{\nu}{a}\frac{\partial\eta}{\partial x}\right); \quad \sigma_{rx} = \mu\frac{\partial u}{\partial r}, \tag{4.109}$$

where ρ_w denotes the wall density, E the wall modulus, ν is Poisson's ratio, h is the wall thickness, and (ξ, η) are the axial and radial displacements respectively. Further,

$$\left.\frac{\partial\xi}{\partial t}\right|_{r=a} = u; \quad \left.\frac{\partial\eta}{\partial t}\right|_{r=a} = v. \tag{4.110}$$

The waveforms for axial and radial flow velocities, pressure, and axial and radial wall velocities are represented by:

$$u(x,r,t) = u_0(r)e^{i(kx-\omega t)},$$

$$v(x,r,t) = v_0(r)e^{i(kx-\omega t)},$$

$$p(x,t) = p_0 e^{i(kx-\omega t)},$$

$$\xi(x,t) = \xi_0 e^{i(kx-\omega t)},$$

$$\eta(x,t) = \eta_0 e^{i(kx-\omega t)}, \tag{4.111}$$

where ω ($= 2\pi/T$) is the frequency, λ ($= 2\pi/k$) is the wave length and ω/k is the wave speed denoted by c.

The actual waveforms will be a composite of harmonic components with different frequencies. However, for the purpose of analysis, it suffices to concern ourselves with a pulse wave of frequency ω, obtain the relationship between the wave amplitudes of the averaged flow velocity and pressure, and thereby demonstrate how a composite pressure wave can be reconstructed from the composite flow velocity wave, which can be monitored by means of pulsed doppler flowmetry.

Substituting the expressions for u and p in Eq. (4.106), we get the following non-homogeneous Bessel equation in u_0:

$$\frac{d^2u_0}{dr^2} + \frac{1}{r}\frac{du_0}{dr} + \frac{i\omega\rho}{\mu}u_0 = \frac{ikp_0}{\mu}, \tag{4.112}$$

whose solution is given by

$$u_0(r) = \frac{k}{\omega\rho}p_0 + AJ_0(\beta r) + BY_0(\beta r); \quad \beta = \frac{i\omega\rho}{\mu}, \tag{4.113}$$

in which the term $Y_0(\beta r)$ becomes irregular at $r = 0$, and hence, must be omitted.

Likewise, we substitute the waveform expressions for u and v in Eq. (4.105), integrate the resulting differential equation:

$$\frac{d(rv_0)}{dr} = -ik(ru_0), \tag{4.114}$$

substitute for $u_0(r)$ from Eq. (4.113), and by keeping in mind that $v_0|_{r=0} = 0$, (by virtue of axial symmetry) and that $\int rJ_0(r)dr = rJ_1$, we obtain:

$$v_0(r) = -\frac{ik^2}{2\omega\rho}p_0r - \frac{ikA}{\beta}J_1(\beta r). \tag{4.115}$$

Next, we substitute the waveform expressions (4.111) in the dynamic equations (4.108) and (4.109) of the tube wall segment, and obtain,

$$p_0 = \frac{Eh}{(1-\nu^2)a^2}\eta_0 + \frac{iEhk\nu}{(1-\nu^2)a}\xi_0 - \rho_w h\omega^2\eta_0, \tag{4.116}$$

$$\xi_0 = \frac{i\nu}{ak}\eta_0 + \frac{(1-\nu^2)\mu A\beta}{Ehk^2}J_1(\beta a) - \frac{\rho_w h\omega^2(1-\nu^2)}{Ehk^2}\xi_0. \tag{4.117}$$

The last terms in Eqs. (4.116) and (4.117) are of relatively small order of magnitude, and may thus be neglected.

Next by invoking the boundary conditions (4.110):

$$-i\omega\xi_0 = u_0(a); \quad -i\omega\eta_0 = v_0(a), \tag{4.118}$$

wherein $u_0(a)$ and $v_0(a)$ are given by (4.113) and (4.115), so that

$$\xi_0 = \frac{-k}{i\omega^2\rho}p_0 - \frac{A}{i\omega}J_0(\beta a), \qquad (4.119)$$

$$\eta_0 = \frac{k^2 a p_0}{2\omega^2\rho} + \frac{Ak}{\beta\omega}J_1(\beta a), \qquad (4.120)$$

and by substituting the above equations into Eqs. (4.116) and (4.117), we obtain:

$$p_0\left(\frac{E^*}{\rho}\frac{k^2}{\omega^2}\left(\frac{1}{2} - \nu\right) - 1\right) + AE^*\frac{k}{\omega}\left(\frac{J_1(\beta a)}{a\beta} - \nu J_0(\beta a)\right) = 0, \qquad (4.121)$$

and

$$p_0\left(\frac{k}{2\omega(i\omega)\rho}(2-\nu)\right) + \frac{A}{\omega^2}\left(\left(\frac{\mu\beta\omega^2}{E^*ak^2} + \frac{i\omega\nu}{a\beta}\right)J_1(\beta a) + \omega\frac{J_0(\beta a)}{i}\right) = 0, \quad (4.122)$$

wherein $E^* = Eh/(1 - \nu^2)a$.

For a non-zero wave amplitude, the determinant of the above set of linear equations in p_0 and A must be zero, which finally yields the following characteristic equation in $\overline{X} = \frac{k^2}{\omega^2}E^*$, after some mathematical simplification:

$$((1 - 2\nu)\overline{X} - 2\rho)\left(\left(\frac{\nu}{a\beta} - S\right)\overline{X} + \frac{\rho}{a}\right) + \frac{(2-\nu)}{2\rho}\left(\frac{1}{a\beta} - S\nu\right)\overline{X}^2 = 0, \quad (4.123)$$

where

$$S = \frac{J_0(\beta a)}{J_1(\beta a)}. \qquad (4.124)$$

From this equation we can determine the value of the parameter $\overline{X}$ in terms of the known values a, ν, S and β. The above analysis will now be applied to determine the cardiac output (CO).

Now ω/k, the wave speed, can be evaluated by monitoring the transit time as the time interval between the peaks or centroids of ultrasonically measured waveforms of the arterial diameter at two arterial sites at a known distance apart. Then, since the value of $\overline{X}$ can be determined from Eq. (4.123), that of E^* can also be calculated.

Having evaluated E^* and ω/k, we can now compute the value of A in terms of p_0 from either Eq. (4.121) or (4.122) as,

$$A = \frac{-\left(\frac{E^*k^2}{2\rho\omega^2}(1 - 2\nu) - 1\right)p_0}{\frac{E^*k}{\omega}\left(\frac{J_1(\beta a)}{a\beta} - \nu J_0(\beta a)\right)}, \qquad (4.125)$$

and thereby express $u_0(r)$ in Eq. (4.113) completely in terms of p_0 as follows:

$$u_0(r) = \left(\frac{k}{\omega\rho} - \theta J_0(\beta r)\right)p_0, \qquad (4.126)$$

where

$$\theta = \frac{\left(\frac{E^* k^2 (1-2\nu)}{2\rho\omega^2} - 1\right)}{E^* \frac{k}{\omega} \left(\frac{J_1(\beta a)}{a\beta} - \nu J_0(\beta a)\right)}. \tag{4.127}$$

The above equation enables the lumen averaged velocity amplitude, $\bar{u}_0$ (corresponding to the pulse frequency ω), given by

$$\bar{u}_0 = \frac{1}{\pi a^2} \int_0^a u_0(r) 2\pi r \, dr, \tag{4.128}$$

to be related to the corresponding pressure amplitude, p_0, in terms of quantities that are known (such as ρ), monitorable (such as a), and computable (such as S and β). We can now adopt the following steps to construct the arterial pressure waveform:

(i) By means of pulsed Doppler flowmetry, monitor (at an arterial site $x = x_0$), the lumen averaged flow velocity waveform, namely

$$\bar{u}(x_0, t) = \sum_{n=-\infty}^{\infty} \bar{u}_n e^{in\omega t}. \tag{4.129}$$

(ii) Determine the spectral components, given by,

$$\bar{u}_n = \frac{1}{T} \int_0^T \bar{u}(t) e^{-in\omega t} \, dt. \tag{4.130}$$

(iii) For each harmonic component $\bar{u}_n$ with frequency $\omega_n = n\omega$, we can determine the associated p_n from Eqs. (4.126) and (4.128), as

$$\begin{aligned} p_n &= \frac{\pi a^2 \bar{u}_n}{\int_0^a \left(\frac{k}{\omega_n \rho} - \theta J_0(\beta r)\right) 2\pi r \, dr} \\ &= \frac{a \bar{u}_n}{\left(\frac{ak}{\omega_n \rho} - \frac{2\theta}{\beta} J_1(\beta a)\right)}. \end{aligned} \tag{4.131}$$

(iv) Obtain the pressure waveform (at a certain $x = x_0$), which is given by,

$$p(x_0, t) = \sum_{n=-\infty}^{\infty} p_n e^{i\omega_n t}. \tag{4.132}$$

(v) Finally calibrate $p(x_0, t)$ by making the peak cyclic value equals to the systolic arterial pressure value obtained by means of sphygmomanometry, which amounts to shifting the $p(x_0, t)$ waveform upward by its mean value amount.

We can also express the ascending aortic flow rate, Q, and hence, the cardic ouput, CO, in terms of p_0, k and ω, i.e., in terms of the parameters of the ascending aortic pressure waveform.

The cardiac output is given by,

$$CO = \int_0^T Q \, dt, \tag{4.133}$$

where

$$Q(t) = \pi a^2 \bar{u}(x_0, t);$$

$$\bar{u}(x_0, t) = \sum_{n=-\infty}^{\infty} \bar{u}_n e^{in\omega t};$$

$$\bar{u}_n = \frac{1}{\pi a^2} \int_0^a u_n(r) 2\pi r \, dr;$$

$$u_n(r) = \left(\frac{k}{\omega_n \rho} - \theta J_0(\beta r) \right) p_n; \quad \omega_n = n\omega,$$

$$p_n = \frac{1}{T} \int_0^T p(x_0, t) e^{-i\omega_n t} \, dt. \tag{4.134}$$

Hence, the cardiac output can be calculated knowing the values of the measurable parameters.

4.11 Flow Through a Converging-Diverging Duct

We will now study the flow of an incompressible fluid through an axisymmetric converging-diverging tube. The analysis will be carried out in the context of a mathematical model for mild *stenosis*.

In the arterial systems of humans or animals, it is quite common to find localised narrowings, commonly called stenoses, caused by intravascular plaques. These stenoses disturb the normal pattern of blood flow through the artery. A knowledge of the flow characteristics in the vicinity of a stenosis may help to further understand some major complications which can arise such as, an ingrowth of tissue in the artery, the development of a coronary thrombosis, the weakening and bulging of the artery downstream from the stenosis, etc.

Investigation of the role of hydrodynamic factors in the development and progression of the above complications provides relevance to this analysis of flow through a model arterial stenosis. Conversely, the expression and resulting nomograms for the increased mean pressure drop through a stenotic segment (averaged over the period of the heart cycle, T), in terms of the vessel size and cardiac output, can provide us an index of the degree of stenosis.

It has been observed that at certain arterial sites, stagnation zones occur, which are due to the separation of the main flow from the walls of the arteries. It has been suggested that stagnation zones near prosthetic heart valves contribute to the formation of a thrombus. Thus, it seems reasonable to speculate that if separation regions of relatively stagnant flow occur in stenotic segments, they may well contribute to thrombosis.

The relationship between the mean pressure, p_0, averaged over the cardiac cycle period, T, (whereby $p_0 = (1/T) \int_0^T p\, dt$) and mean flow, q_0 ($= (1/T) \int_0^T q\, dt$), can be modelled as the Poiseuille (steady) flow of a viscous and incompressible Newtonian fluid (of viscosity μ, taken as 0.04 Poise, or dyne sec cm^{-1}, for blood at 37°C) through a straight circular rigid tube of radius a. The familiar relationship between q_0 and the gradient of p_0 is given by,

$$q_0 = -\frac{\pi a^4}{8\mu}\frac{dp_0}{dx}. \tag{4.135}$$

The associated axial velocity profile is parabolic and given by,

$$u(r) = -\frac{1}{4\mu}\frac{dp_0}{dx}(a^2 - r^2). \tag{4.136}$$

However, if a converging-diverging segment is placed in a straight tube, the velocity profile will be altered, with a subsequent change in other flow characteristics. In order to assess the mean pressure drop due to a stenosis of measurable geometry under a pulsed Doppler based monitorable mean (lumen averaged) flow velocity, the specific problem considered here will be that of steady axisymmetric flow of an incompressible fluid through a converging-diverging tube with circular cross-section having rigid walls (Forrester and Young [25]; Young [127]).

Figure 4.12 illustrates the geometry of a collar-like stenosis located in a relatively straight artery. Consider that the axial coordinate and velocity are to be denoted as $\hat{z}$ and $\hat{u}$ respectively and the radial coordinate and velocity are to be similarly denoted as $\hat{r}$ and $\hat{v}$ respectively. The local radius of the axisymmetric tube is $\hat{R}$, whereas R_0 is the radius of the artery proximal to the stenoised segment.

Fig. 4.12: Collar-like stenosis in a relatively straight artery.

The governing equations are the continuity equation and the Navier-Stokes equations, wherein we will first introduce the following dimensionless variables:

$$r = \frac{\hat{r}}{R_0}, \quad z = \frac{\hat{z}}{R_0}, \quad R = \frac{\hat{R}}{R_0}, \quad u = \frac{\hat{u}}{\overline{U}_0},$$
$$\left. \begin{array}{cc} & \\ v = \frac{\hat{v}}{\overline{U}_0}, \quad p = \frac{\hat{p}}{\rho \overline{U}_0^2}, \quad U = \frac{\hat{U}}{\overline{U}_0}, \end{array} \right\} \qquad (4.137)$$

where $\hat{U}$ is the centreline velocity, $\overline{U}_0$ is the lumen averaged velocity in the unobstructed tube, $\hat{p}$ is the pressure and ρ is the fluid density. It is to be noted when applying the analysis to an arterial segment, that the values of R_0 and $\overline{U}_0$ will be time averaged over a cardiac cycle period.

The continuity equation and the Navier-Stokes equations (in the axial and radial directions) can be written in dimensionless form as follows,

$$\frac{\partial u}{\partial z} + \frac{\partial v}{\partial r} + \frac{v}{r} = 0, \qquad (4.138)$$

$$u \frac{\partial u}{\partial z} + v \frac{\partial u}{\partial r} + \frac{\partial p}{\partial z} = \frac{2}{Re} \left(\frac{\partial^2 u}{\partial r^2} + \frac{1}{r} \frac{\partial u}{\partial r} + \frac{\partial^2 u}{\partial z^2} \right), \qquad (4.139)$$

$$u \frac{\partial v}{\partial z} + v \frac{\partial v}{\partial r} + \frac{\partial p}{\partial r} = \frac{2}{R_e} \left(\frac{\partial^2 v}{\partial r^2} + \frac{1}{r} \frac{\partial v}{\partial r} - \frac{v}{r^2} + \frac{\partial^2 v}{\partial z^2} \right), \qquad (4.140)$$

where R_e denotes the Reynolds number $(2R_0 \overline{U}_0 \rho / \mu)$ and μ denotes the fluid viscosity. These equations are non-linear and exact solutions are not easy to obtain. However, approximate solutions can be obtained on the basis of some simplifying assumptions (Morgan and Young [75]).

Integrating Eq. (4.139) over the cross-section of the tube, and imposing the no-slip boundary condition,

$$u = v = 0$$

at the wall, the integral momentum equation is obtained as,

$$\frac{1}{2} \frac{\partial}{\partial z} \int_0^R r u^2 \, dr = - \int_0^R r \frac{\partial p}{\partial z} \, dr + \frac{2}{R_e} \left(R \frac{\partial u}{\partial r} \Big|_R + \int_0^R r \frac{\partial^2 u}{\partial z^2} \, dr \right). \qquad (4.141)$$

Likewise, by first multiplying Eq. (4.139) by ru, and then integrating from $r = 0$ to $r = R$, we obtain,

$$\frac{1}{3} \frac{\partial}{\partial z} \int_0^R r u^3 \, dr = - \int_0^R r u \frac{\partial p}{\partial z} \, dr + \frac{2}{R_e} \left(- \int_0^R r \left(\frac{\partial u}{\partial r} \right)^2 \, dr + \int_0^R r u \frac{\partial^2 u}{\partial z^2} \, dr \right). \qquad (4.142)$$

In Eqs. (4.141) and (4.142), the terms containing $\partial^2 u/\partial z^2$ associated with the viscous component of the normal stress in the axial direction are considered negligible. Further, the $\int_0^R r \frac{\partial p}{\partial z} \, dr$ term appearing in Eq. (4.141) can be reasonably approximated as

$$\int_0^R r \frac{\partial p}{\partial z} \, dr \approx R^2 \int_0^R r u \frac{\partial p}{\partial z} \, dr, \qquad (4.143)$$

assuming in a constriction the pressure gradient is nearly constant and the velocity profile tends to become flattened.

Invoking these assumptions, Eqs. (4.141) and (4.142) are combined into a single equation in terms of the non-dimensional axial velocity u:

$$\frac{1}{3} R^2 \frac{\partial}{\partial z} \int_0^R r u^3 \, dr - \frac{1}{2} \frac{\partial}{\partial z} \int_0^R r u^2 \, dr \; = \; -\frac{2}{R_e} \left(R^2 \int_0^R r \left(\frac{\partial u}{\partial r} \right)^2 dr + R \left. \frac{\partial u}{\partial r} \right|_R \right).$$
$$(4.144)$$

At this stage, we need to assume the axial velocity profile function, in order to fulfill the following characteristics of flow through the converging-diverging cross-section:

(i)   At high Reynolds number, the profile must yield a thin region of high shear near the wall in the converging section, with a flat profile in the core.

(ii) The profile must assume the characteristics of a central jet with low shear, with backflow near the wall, in the diverging section.

(iii) The no slip condition,
$$u = 0 \quad \text{at} \quad r = R.$$

(iv) At the centreline,
$$u = U \quad \text{at} \quad r = 0.$$

(v)   The smoothness condition associated with zero stress jump at the origin,
$$\frac{\partial u}{\partial r} = 0 \quad \text{at} \quad r = 0.$$

(vi) By eliminating the pressure term between Eqs. (4.139) and (4.140), and considering the resulting equation in the limit as r approaches zero, yields
$$\frac{\partial^3 u}{\partial r^3} = 0 \quad \text{at} \quad r = 0.$$

(vii) In order that the net flow through any cross-section be a constant,
$$\int_0^R r u \, dr \; = \; \frac{1}{2}.$$

The velocity profile that satisfies the above requirements is given in two parts:

(a)

$$u = U\left(A + B\left(\frac{r}{R}\right) + C\left(\frac{r}{R}\right)^2 + D\left(\frac{r}{R}\right)^3 + E\left(\frac{r}{R}\right)^4\right) \tag{4.145}$$

and the unknown coefficients are obtained by satisfying the above constraints (iii)–(vii) to give,

$$u = \frac{1}{R^2}\left\{R^2U + 2(3 - 2R^2U)\left(\frac{r}{R}\right)^2 - 3(2 - R^2U)\left(\frac{r}{R}\right)^4\right\}, \text{ for } R^2U \geq 1.5. \tag{4.146}$$

Thus, the velocity profile is determined in terms of centreline velocity U. It is interesting to note that when $R^2U = 2$, Eq. (4.146) reduces to $u = \frac{2}{R^2}[1-(r/R)^2]$, which is a parabolic profile corresponding to the Poiseuille flow.

(b)

$$u = \begin{cases} U, & \text{for } 0 \leq \left(\frac{r}{R}\right) \leq \lambda; \\ a + b\left(\frac{r}{R}\right)^2 + c\left(\frac{r}{R}\right)^4, & \text{for } \lambda < \frac{r}{R} \leq 1, \end{cases} \tag{4.147}$$

which by satisfying the no-slip condition, and $u = U$ and $\frac{\partial u}{\partial r}$ at $r/R = \lambda$, yields

$$u = \begin{cases} U, & \text{for } 0 \leq \left(\frac{r}{R}\right) \leq \lambda; \\ \dfrac{U}{(1 - \lambda^2)^2}\left(1 - 2\lambda^2 + 2\lambda^2\left(\frac{r}{R}\right)^2 - \left(\frac{r}{R}\right)^4\right), & \text{for } \lambda < \left(\frac{r}{R}\right) \leq 1, \end{cases} \tag{4.148}$$

with

$$\lambda = \left(\frac{3}{R^2U} - 2\right)^{1/2}, \quad \text{and for } R^2U \leq 1.5.$$

The resulting profiles for various values of R^2U are shown in Fig. 4.13.

Upon substituting the profiles (4.146) and (4.148) into Eq. (4.144), we obtain

$$\frac{dU}{dz} = \left\{-\frac{1}{R^3}\frac{dR}{dz}\left(0.0286(R^2U)^3 - 0.133(R^2U)^2 + 0.257R^2U - 0.34\right)\right.$$

$$\left. + \frac{4U\Gamma}{Re}\right\}\left(0.0429(R^2U)^2 - 0.004(R^2U) - 0.029\right)^{-1}, \tag{4.149}$$

where

$$\Gamma = \begin{cases} 2 - R^2U, & \text{for } R^2U \geq 1.5; \\ \dfrac{3R^2U - 1}{9(R^2U - 1)}, & \text{for } R^2U < 1.5. \end{cases} \tag{4.150}$$

The initial condition required to numerically solve this nonlinear differential equation

Fig. 4.13: Velocity profile shapes.

is,

$$\text{at } z = -\infty, \quad R = 1 \text{ and } U = 2.$$

What is of interest to us in being able to lead to a noninvasive diagnostic method for detecting the location and degree of stenosis, is the expression for the average pressure gradient given by

$$\left.\frac{\partial p}{\partial z}\right|_{av} = \frac{\int_0^R r \frac{\partial p}{\partial z}\, dr}{\int_0^R r\, dr} = \frac{2}{R^2}\int_0^R r\frac{\partial p}{\partial z}\, dr \qquad (4.151)$$

wherein, from Eq. (4.141)

(i)

$$\int_0^R r\frac{\partial p}{\partial z}\, dr = \frac{2}{Re}\left(\left.R\frac{\partial u}{\partial r}\right|_R + \int_0^R r\frac{\partial^2 u}{\partial z^2}\, dr\right) - \frac{1}{2}\frac{\partial}{\partial z}\int_0^R ru^2\, dr \qquad (4.152)$$

(ii) u is given, by Eqs. (4.146) and (4.148), in terms of U.

(iii) U is solved by means of Eq. (4.149), for stenosis geometry such as that given by

$$R(z) = \begin{cases} 1 - \frac{\delta}{2}\left(1 + \cos\left(\frac{\pi z}{Z_0}\right)\right), & \text{for } |z| \le Z_0; \\ 1, & \text{for } |z| > Z_0, \end{cases} \qquad (4.153)$$

where δ is the (stenosis or plaque height)$/R_0$ and Z_0 is the (stenosis length)$/2R_0$.

By means of Eqs. (4.146)–(4.150) and (4.153) the non-dimensional pressure drop Δp can be numerically computed and plotted as a function of δ, Z_0 and R_0. For this purpose using a set of values of (δ, Z_0, R_0) we numerically formulate the shape function $R(z)$ as a function of the non-dimensional coordinate z, solve Eq. (4.149) for U as a function of z, substitute the values of $U(z)$ in Eqs. (4.146) and (4.148) and formulate u as a function of r and z, then employ Eqs. (4.151) and (4.152) to compute the values of $\partial p/\partial z$ for different values of z, and finally determine the value of the pressure drop,

$$\Delta p = \int_{-Z_0}^{Z_0} \frac{\partial p}{\partial z}\, dz. \qquad (4.154)$$

Thus, from the computed nomograms (Fig. 4.14), the non-dimensional pressure drop Δp can be determined for known values of Z_0 (the non-dimensional distance between arterial measurement sites proximal and distal to the stenotic site), δ (the non-dimensional parameter for the degree of stenosis) and R_0 (the radius of the unconstrained sections upstream and downstream of the stenotic site).

Consider now a possible methodology for practical implementation of this analysis, to noninvasively determine the pressure drop Δp across the stenotic segment. By means of a pulse Doppler flowmeter, we can determine the lumen averaged instantaneous flow velocity proximal to the stenotic segment in terms of the mean Doppler frequency shift (Baker and Daigle [6]), as

$$\overline{U} = \frac{c\Delta f}{2f_0 \cos\theta}; \qquad (4.155)$$

wherein Δf is the mean Doppler frequency shift (averaged over the vessel lumen), c the velocity of sound, f_0 the centre frequency and θ the angle between the velocity and beam vectors. Therefore we can determine the averaged flow velocity $\overline{U_0}$ over a period T, as

$$\overline{U}_0 = \frac{1}{T} \int_0^T \overline{U}(t)\, dt. \qquad (4.156)$$

We then obtain real-time longitudinal cross-sectional views of the stenotic segment by means of a pulse echo sector-scan unit, measure from there the value of the stenosis length, Z_0, as well as the instantaneous values of the stenosis height and the unconstrained tube radius, average them over the cardiac period to obtain the averaged values of R_0 and the stenosis height parameter δ, as defined by Eq. (4.153). Having measured the values of δ, R_0 and Z_0, we can then obtain the value of Δp from the developed nomograms of "Δp vs δ and Z_0", of Fig. 4.14.

(a) Plot of Δp vs Z_0 (b) Plot of Δp vs δ

Fig. 4.14: Plots of Δp.

Chapter 5

NON-NEWTONIAN FLUIDS

5.1 General Introduction

The study of fluid dynamics began with imaginary "ideal" or "perfect" fluid that is incompressible and without viscosity or elasticity. Shearing stresses are absent during shear as the fluid movement is completely frictionless. Detailed mathematical relationships have been obtained for the behaviour of a perfect fluid in a wide variety of physical situations. Some of these results have proved to be useful approximations to the performance of real fluids in certain special cases.

However, severe limitations in the practical application of frictionless flow theory to real situations in general led to the development of a dynamical theory for the simplest class of real fluids — Newtonian fluids.

A Newtonian fluid, by definition, is one in which the coefficient of viscosity is constant at all rates of shear. Homogeneous liquids may behave closely like Newtonian fluids. However, there are fluids that do not obey the linear relationship between shear stress and shear strain rate. Fluids that exhibit a non-linear relationship between the shear stress and the rate of shear strain are called *non-Newtonian* fluids. Thick liquids are generally in this category. Many common fluids are non-Newtonian, e.g. paints, enamels, varnish, wet clay and mud, solutions of various polymers, suspensions of particles, emulsions of oil in water, etc.

Although the properties of non-Newtonian fluids do not lend themselves to precise and simple analysis that has been developed for Newtonian fluids, the flow of non-Newtonian fluids does possess some interesting and useful characteristics.

We have seen that blood deviates in behaviour from Newtonian fluids. Such anomalous behaviour of blood is often referred to as non-Newtonian properties of blood. These properties are of two types as follows:

(i) at low shear rates, the apparent viscosity increases markedly — sometimes even a certain "yield stress" is required for flow.

(ii) in small tubes, the apparent viscosity at higher rates of shear is smaller than it is in larger tubes.

These two types of anomalies are often referred to as "low shear" and "high shear" effects respectively. It is thus concluded that the behaviour of blood is almost Newtonian at high shear rate, while at low shear rate the blood exhibits yield stress and non-Newtonian behaviour. Most body fluids are non-Newtonian.

5.2 Classification of Non-Newtonian Fluids

Fluids which do not obey the linear relationship between shear stress and rate of shear strain can be grouped into three general classifications:

(1) The simplest type of non-Newtonian fluids is the time independent non-Newtonian fluid in which the shear strain rate is a non-linear function of the shear stress, independent of shearing time and previous shear stress rate history of that fluid.

(2) Time dependent non-Newtonian fluids have more complex shearing stress strain rate relationships. The shearing strain rates in these fluids depend on shearing times or on the previous shear stress rate histories of the fluids and, thus, are not single valued functions of the shear stresses.

(3) In *viscoelastic fluids*, some of the energy of deformation may be recoverable as it is in elastic solids. This is different from truly viscous fluids in which all the energy of deformation is dissipated. Hence, the shear strain as well as shear strain rate are related in some way to shear stress.

5.3 Time Independent Fluids

For a time independent non-Newtonian fluid, the constitutive equation is

$$\tau = f(\dot{\gamma})$$

or

$$\dot{\gamma} = f(\tau). \tag{5.1}$$

A Newtonian fluid is just a special case of non-Newtonian fluid where the function

$f(\dot{\gamma})$ is linear. Figure 5.1 shows some typical shear stress strain rate relations for non-Newtonian fluids.

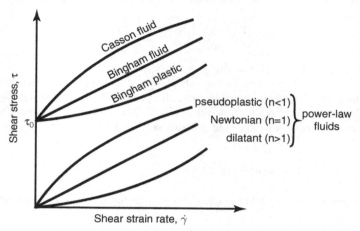

Fig. 5.1: Typical shear stress strain rate relationships for non-Newtonian fluids.

5.3.1 *Power-Law Fluids*

One important class of non-Newtonian fluids is that of power-law fluids which have constitutive equation

$$\tau = \mu\dot{\gamma}^n$$
$$= \mu\dot{\gamma}^{n-1}\dot{\gamma}. \tag{5.2}$$

This class of non-Newtonian fluids has effective viscosity coefficient or apparent viscosity $\mu\dot{\gamma}^{n-1}$ and does not have a yield stress. If $n < 1$, we get a pseudoplastic power-law fluid characterised by a progressively decreasing apparent viscosity with strain rate. If $n > 1$, we have a dilatant power-law fluid in which the apparent viscosity increases progressively with increasing strain rate. If $n = 1$, we obtain the Newtonian fluid as a special case. At very high shear rates in a real fluid, the apparent viscosity becomes constant and equal to μ_∞, and the shear stress shear rate relationship becomes linear.

Several objections are raised against the power-law model, one of them being for a pseudoplastic fluid, the apparent viscosity is infinite when the shear strain rate is zero. However, this model has been found desirable because of its simplicity, as well as being adequate to analyse such flows as *Couette flow* and flow in pipes and channels. We will discuss this later.

5.3.2 *Bingham Fluids*

Another important non-Newtonian fluid, the Bingham fluid, has constitutive equation

$$\tau = \mu\dot{\gamma} + \tau_0, \quad \tau \geq \tau_0$$
$$\dot{\gamma} = 0. \quad \tau \leq \tau_0. \tag{5.3}$$

It is also known as Bingham plastic.

It exhibits a yield stress τ_0 at zero shear rate, followed by a linear relationship between shear stress and shear strain rate, the plastic viscosity μ being the slope of the straight line. If $\tau < \tau_0$, no flow takes place. The behaviour of many real fluids such as slurries, household paint and plastics very closely approximate this concept.

5.3.3 *Other Special Non-Newtonian Fluids*

The following are some other empirical equations for special non-Newtonian fluids:

Herschel-Bulkley fluid:

$$\tau = \mu\dot{\gamma}^n + \tau_0, \quad \tau \geq \tau_0$$
$$\dot{\gamma} = 0, \quad \tau \leq \tau_0 \tag{5.4}$$

Casson fluid:

$$\tau^{\frac{1}{2}} = \mu^{\frac{1}{2}}\dot{\gamma}^{\frac{1}{2}} + \tau_0^{\frac{1}{2}}, \quad \tau \geq \tau_0$$
$$\dot{\gamma} = 0, \quad \tau \leq \tau_0 \tag{5.5}$$

Prandtl fluid:

$$\tau = A \arcsin\left(\frac{\dot{\gamma}}{C}\right) \tag{5.6}$$

Prandtl-Eyring fluid:

$$\tau = A\dot{\gamma} + B \arcsin\left(\frac{\dot{\gamma}}{C}\right) \tag{5.7}$$

Power-Eyring fluid:

$$\tau = A\dot{\gamma} + B \sinh^{-1}(C\dot{\gamma}) \tag{5.8}$$

Ellis fluid:

$$\dot{\gamma} = A\tau + B\tau^n \tag{5.9}$$

Reiner-Philliphoff fluid:

$$\tau = \left(\frac{\mu_0 - \mu_\infty}{1 + \left(\frac{\tau}{\tau_0}\right)^2}\right)\dot{\gamma} + \mu_\infty \tag{5.10}$$

Rabinowitsch fluid:

$$\dot{\gamma} = \frac{1}{\mu_0}\tau + \sum_q s_q \tau^{2q+1}. \tag{5.11}$$

The above list of non-Newtonian fluids is by no means complete. However, for biofluid flows, Eqs. (5.2), (5.4) and (5.5) are especially useful.

5.4 Time Dependent Fluids

Some fluids are more complex than those just described and the apparent viscosity depends not only on the strain rate but also on the time the shear has been applied. These can generally be classified into two classes:

(i) thixotropic fluids

(ii) rheopectic fluids.

For a thixotropic fluid, the shear stress decreases with time as the fluid is sheared while for a rheopectic fluid, the shear stress increases with time as the fluid is sheared. An example of a thixotropic fluid is printer's ink.

5.5 Viscoelastic Fluids

A viscoelastic material exhibits both elastic and viscous properties. The simplest viscoelastic fluid is one which is Newtonian in viscosity and obeys Hooke's law for the elastic part, giving the constitutive equation

$$\dot{\gamma} = \frac{\tau}{\mu_0} + \frac{\dot{\tau}}{\lambda} \tag{5.12}$$

where λ is a rigidity modulus.

Rather complex models of viscoelastic materials have been developed in which higher time derivatives of τ and γ appear. For time varying processes, the elastic constants may be complex functions of frequency.

5.6 Laminar Flow of Non-Newtonian Fluids

Let us now discuss a number of non-Newtonian fluid models for laminar flow in a tube and comment on how well they compare to experimental results. The assumptions of no external forces and steady, uniaxial, axisymmetric flow in a rigid, cylindrical tube still apply. It is further assumed that the flow is fully-developed far away from the ends of the tube.

5.6.1 *Power-Law Model*

A simple model taking into account the non-Newtonian behaviour of fluid flow is the power-law model. It suggests a constitutive equation of the form

$$\tau = \mu\dot{\gamma}^n \qquad\qquad (5.13)$$

where τ is the shear stress
  $\dot{\gamma}$ is the shear strain rate.

At this point, since we are dealing with variable viscosity, we shall define what we mean by "variable". If the viscosity is constant, as in the Newtonian case, then it is simply the ratio of stress (τ) to shear rate ($\dot{\gamma}$). However, for variable viscosity, we have "differential" viscosity μ' being defined by the slope of the curve, i.e.,

$$\mu' = \tan^{-1}\left(\frac{d\tau}{d\dot{\gamma}}\right).$$

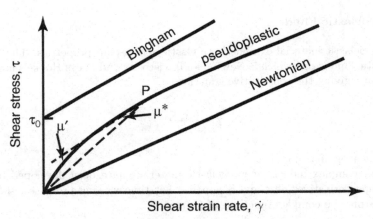

Fig. 5.2: Differential and generalised viscosity for power-law model.

Alternatively, we may consider "generalised" viscosity μ^*, which is the ratio of applied stress to shear rate (see Fig. 5.2). Thus, at point P,

$$\mu^* = \tan^{-1}\left(\frac{\tau}{\dot{\gamma}}\right).$$

This is also called "secant viscosity".

Let us first determine the flux of the flow. Consider a control volume bounded by two coaxial cylinders of radii r and $r + dr$ and both of length L (see Fig. 5.3).

Fig. 5.3: The control volume and the forces acting on its surfaces.

Basically, two types of forces act on this control volume; forces due to the pressure gradient and forces due to the shear stresses.

The force due to a constant pressure gradient of $-P$ across a cylindrical tube of length L is

$$F_p = P \times 2\pi r L \, dr. \tag{5.14}$$

An object under shear stress has equal and opposite forces applied across its opposite faces. Thus there are shear stresses acting both on the inner and outer cylindrical surfaces of the control volume. Let the stress at a distance r from the axis be denoted by, $\tau(r)$, and at a distance $(r + dr)$ be denoted by $\tau(r + dr)$. Then the force acting on the inner cylindrical surface of the volume will be

$$F_{is} = 2\pi r L \tau(r) \tag{5.15}$$

and that on the outer cylindrical surface is

$$
\begin{aligned}
F_{os} &= 2\pi(r + dr)L\tau(r + dr), \quad \text{which after using Taylor series expansion} \\
&= 2\pi(r + dr)L\left[\tau + dr\frac{d\tau}{dr} + O\left((dr)^2\right)\right] \\
&= 2\pi L\left[r\tau + \tau dr + r\,dr\frac{d\tau}{dr} + O\left((dr)^2\right)\right] \\
&\simeq 2\pi L\left[r\tau + \frac{d}{dr}(r\tau)\,dr\right].
\end{aligned}
\tag{5.16}
$$

Balancing forces parallel to the axis,

$$F_{os} - F_{is} = F_p$$

$$2\pi L \, dr \frac{d}{dr}(r\tau) = P \times 2\pi r L \, dr$$

$$\frac{d}{dr}(r\tau) = Pr. \tag{5.17}$$

Integrating Eq. (5.17), we have

$$\tau = \frac{Pr}{2} + \frac{A}{r}. \tag{5.18}$$

For stress to be finite on the axis ($r = 0$), we need $A = 0$. Therefore,

$$\tau = \frac{Pr}{2}. \tag{5.19}$$

As the velocity is a function of r only, the only non-zero component of the strain rate tensor is

$$\dot{\gamma} = -\frac{dv}{dr}. \tag{5.20}$$

If we consider a time-independent non-Newtonian fluid, for which the constitutive equation is given by Eq. (5.1), we obtain

$$\frac{Pr}{2} = f\left(-\frac{dv}{dr}\right). \tag{5.21}$$

The flux Q is given by

$$Q = \int_0^R 2\pi r v \, dr \tag{5.22}$$

which after integration by parts and invoking the no-slip boundary condition that $v = 0$ when $r = R$, reduces to

$$Q = \pi \int_0^R r^2 \left(-\frac{dv}{dr}\right) dr. \tag{5.23}$$

Thus, knowing the precise form of the functional relation given by Eq. (5.21), the flux Q can be obtained.

In the case of the power-law model, the constitutive equation is given by Eq. (5.13). Equating Eqs. (5.13) and (5.19), we obtain

$$\dot{\gamma} = \left(\frac{Pr}{2\mu}\right)^{\frac{1}{n}}. \tag{5.24}$$

Using Eq. (5.20), the above equation reduces to

$$\frac{dv}{dr} = -\left(\frac{Pr}{2\mu}\right)^{\frac{1}{n}}. \tag{5.25}$$

Integrating this, we obtain

$$v = -\left(\frac{P}{2\mu}\right)^{\frac{1}{n}} \frac{n}{n+1} r^{1+\frac{1}{n}} + c. \tag{5.26}$$

The no-slip condition implies that

$$v = \left(\frac{P}{2\mu}\right)^{\frac{1}{n}} \frac{n}{n+1}(R^{1+\frac{1}{n}} - r^{1+\frac{1}{n}}). \tag{5.27}$$

Using Eqs. (5.23) and (5.25), the flux Q is given by

$$Q = \pi \int_0^R r^2 \left(\frac{Pr}{2\mu}\right)^{\frac{1}{n}} dr$$

$$= \left(\frac{P}{2\mu}\right)^{\frac{1}{n}} \frac{n\pi}{3n+1} R^{3+\frac{1}{n}}. \tag{5.28}$$

This equation is found to be good for values of n between 0.68 and 0.8 and for strain rates between 5 and 200 sec^{-1}. Note that when $n = 1$, the model reduces to the Newtonian case and the above equation becomes the Poiseuille's equation.

The power-law model is an improvement to the Poiseuille's equation by generalizing the flow to include non-Newtonian effect. The beauty of this model is its simplicity. However, there are other aspects of the flow properties of fluid which this model fails to examine, one of them being the presence of yield stress.

5.6.2 *Herschel-Bulkley Model*

This model includes the yield stress, τ_0, in its constitutive equation, which is given by Eq. (5.4) as

$$\tau = \mu\dot{\gamma}^n + \tau_0, \quad \tau \geq \tau_0$$
$$\dot{\gamma} = 0, \quad \tau \leq \tau_0. \tag{5.29}$$

Let us now determine the flux, Q, due to this model. In this case, due to the existence of a yield stress, there occurs a plug flow. The radius of the plug core region is, say, r_p (see Fig. 5.4). For $r \leq r_p$, $\tau(r) \leq \tau_0$. Thus, for the core region

$$\dot{\gamma} = 0,$$

i.e.,

$$\frac{dv}{dr} = 0. \tag{5.30}$$

Therefore,

$$v = \text{constant}$$

$$= v_p \text{ (say)}. \tag{5.31}$$

Fig. 5.4: Plug flow in the case when a yield stress exists.

Hence, the velocity of flow in the core region is a constant and the velocity profile in this region is straight (see Fig. 5.4).

Considering forces in the core region, we have

$$2\pi r_p L \tau_0 = P \times \pi r_p^2 L.$$

Thus,

$$r_p = \frac{2\tau_0}{P}, \tag{5.32}$$

or

$$\tau_0 = \frac{P r_p}{2}. \tag{5.33}$$

Outside the core region, $\tau \geq \tau_0$. The analysis in this region is similar to that in the power-law model. Using Eqs. (5.19), (5.20) and (5.29), we obtain

$$\frac{dv}{dr} = -\left(\frac{P}{2\mu}\right)^{\frac{1}{n}} (r - r_p)^{\frac{1}{n}}. \tag{5.34}$$

Integrating Eq. (5.34), we get

$$v = -\left(\frac{P}{2\mu}\right)^{\frac{1}{n}} \frac{n}{n+1} (r - r_p)^{1+\frac{1}{n}} + c. \tag{5.35}$$

Using the no-slip condition,

$$v = \left(\frac{P}{2\mu}\right)^{\frac{1}{n}} \frac{n}{n+1} \left[(R - r_p)^{1+\frac{1}{n}} - (r - r_p)^{1+\frac{1}{n}}\right]. \tag{5.36}$$

As $v = v_p$ (plug core velocity) when $r = r_p$, we have

$$v_p = \left(\frac{P}{2\mu}\right)^{\frac{1}{n}} \frac{n}{n+1} (R - r_p)^{1+\frac{1}{n}}. \tag{5.37}$$

Using the value of r_p given by Eq. (5.32) into Eqs. (5.36) and (5.37), we can determine the velocity outside the core region and in the core region respectively. Let us now determine the flux across a section of the tube.

$$
\begin{aligned}
Q &= \pi r_p^2 v_p + \int_{r_p}^{R} 2\pi r v \, dr \\
&= \pi \left(\frac{P}{2\mu}\right)^{\frac{1}{n}} \frac{n}{n+1} \Bigg\{ r_p^2 (R - r_p)^{1+\frac{1}{n}} \\
&\quad + 2 \int_{r_p}^{R} r \left[(R - r_p)^{1+\frac{1}{n}} - (r - r_p)^{1+\frac{1}{n}} \right] dr \Bigg\} \\
&= \pi \left(\frac{P}{2\mu}\right)^{\frac{1}{n}} \frac{n}{n+1} \Bigg\{ r_p^2 (R - r_p)^{1+\frac{1}{n}} + 2\left[\frac{(R^2 - r_p^2)}{2} (R - r_p)^{1+\frac{1}{n}} \right. \\
&\quad \left. - \frac{nR}{2n+1}(R - r_p)^{2+\frac{1}{n}} + \frac{(R - r_p)^{3+\frac{1}{n}}}{(2+\frac{1}{n})(3+\frac{1}{n})} \right] \Bigg\} \\
&= \pi \left(\frac{P}{2\mu}\right)^{\frac{1}{n}} \frac{n}{n+1} \Bigg\{ r_p^2 (R - r_p)^{1+\frac{1}{n}} + 2\left[\frac{R + r_p}{2}(R - r_p)^{2+\frac{1}{n}} \right. \\
&\quad \left. - \frac{nR}{2n+1}(R - r_p)^{2+\frac{1}{n}} - \frac{(R - r_p)^{3+\frac{1}{n}}}{3+\frac{1}{n}} + \frac{(R - r_p)^{3+\frac{1}{n}}}{2+\frac{1}{n}} \right] \Bigg\} \\
&= \pi \left(\frac{P}{2\mu}\right)^{\frac{1}{n}} \frac{n}{n+1} \Bigg\{ r_p^2 (R - r_p)^{1+\frac{1}{n}} + 2\left[\frac{R + r_p}{2}(R - r_p)^{2+\frac{1}{n}} \right. \\
&\quad \left. - \frac{n}{3n+1}(R - r_p)^{3+\frac{1}{n}} - \frac{n}{2n+1} r_p (R - r_p)^{2+\frac{1}{n}} \right] \Bigg\} \\
&= \pi \left(\frac{P}{2\mu}\right)^{\frac{1}{n}} \frac{n}{n+1} R^{3+\frac{1}{n}} \left[c_p^2 (1 - c_p)^{1+\frac{1}{n}} + (1 + c_p)(1 - c_p)^{2+\frac{1}{n}} \right.
\end{aligned}
$$

$$
\left. - \frac{2n}{3n+1}(1 - c_p)^{3+\frac{1}{n}} - \frac{2n}{2n+1} c_p (1 - c_p)^{2+\frac{1}{n}} \right], \tag{5.38}
$$

where

$$
c_p = \frac{r_p}{R}. \tag{5.39}
$$

In the case when there is no yield stress, that is $\tau_0 = 0$, the Herschel-Bulkley model, Eq. (5.29), reduces to the power-law model, Eq. (5.2). Thus, substituting $\tau_0 = 0$, i.e., $c_p = 0$ in Eq. (5.38) we obtain the value of flux for the power-law model as given in Eq. (5.28). Further, if we put $n = 1$, we obtain results for the Newtonian fluids.

Thus, the Herschel-Bulkley is a more general model compared to the power-law model. It not only considers the non-Newtonian behaviour of fluid but recognises the existence of yield stress as well. However, it is found to agree with experimental results over a small range of shear rates.

5.6.3 *Casson Model*

The third model that will be considered is the Casson model. The constitutive equation suggested by Casson [16] is

$$\sqrt{\tau} = \sqrt{\mu\dot{\gamma}} + \sqrt{\tau_0}, \qquad \tau \geq \tau_0$$

$$\dot{\gamma} = 0, \qquad\qquad \tau \leq \tau_0. \tag{5.40}$$

In 1958, Casson, basing his calculations on certain assumptions concerning the magnitude of inter-particle forces and disruptive stresses in chain-like floccules present in varnishes and printing inks, derived this semi-empirical equation to describe their flow behaviour. A year later, Scott Blair [99] found that this description is also suitable for the flow of blood. Since then, there has been various studies made supporting the use of the Casson equation to blood flow.

Similar to the Herschel-Bulkley model, Casson's equation includes a yield stress and therefore exhibits plug flow with radius of the core region r_p such that

$$r_p = \frac{2\tau_0}{P} \tag{5.41}$$

and so

$$\tau_0 = \frac{Pr_p}{2}. \tag{5.42}$$

Hence, as explained in the previous section the velocity profile in the core region, $r < r_p$, is flat. Between r_p and R (outside the core region), Casson's equation is applicable.

We will now proceed to find the velocity of flow outside the core region. Squaring the Casson equation and rearranging, we have

$$\dot{\gamma} = \frac{1}{\mu}(\tau + \tau_0 - 2\sqrt{\tau\tau_0}). \tag{5.43}$$

Using Eq. (5.42), and knowing

$$\dot{\gamma} = -\frac{dv}{dr} \tag{5.44}$$

and

$$\tau = \frac{Pr}{2} \tag{5.45}$$

we finally obtain

$$\frac{dv}{dr} = \frac{P}{2\mu}(2\sqrt{r_p r} - r - r_p).$$ (5.46)

Hence,

$$v = \frac{P}{2\mu}\left(\frac{4}{3}\sqrt{r_p r^3} - \frac{r^2}{2} - r_p r + c\right).$$ (5.47)

Applying the no-slip condition, that is, $v = 0$ at $r = R$,

$$c = -\left(\frac{4}{3}\sqrt{r_p R^3} - \frac{R^2}{2} - r_p R\right).$$ (5.48)

Therefore,

$$v = \frac{P}{2\mu}\left[\frac{4}{3}\sqrt{r_p r^3} - \frac{r^2}{2} - r_p r - \left(\frac{4}{3}\sqrt{r_p R^3} - \frac{R^2}{2} - r_p R\right)\right]$$ (5.49)

or

$$v = \frac{P}{4\mu}\left[(R^2 - r^2) - \frac{8}{3}\sqrt{r_p}(\sqrt{R^3} - \sqrt{r^3}) + 2r_p(R - r)\right], \quad r_p \leq r \leq R.$$ (5.50)

As mentioned before, the Casson model has a constant flow in the core region, giving a straight profile in this region. Let $v = v_p$ at $r = r_p$. Then

$$v_p = \frac{P}{4\mu}\left(R^2 + 2r_p R - \frac{8}{3}\sqrt{r_p R^3} - \frac{1}{3}r_p^2\right)$$

$$= \frac{P}{4\mu}(\sqrt{R} - \sqrt{r_p})^3(\sqrt{R} + \frac{1}{3}\sqrt{r_p}).$$ (5.51)

Equation (5.51) represents velocity distribution for all values of r between 0 and r_p, known as plug velocity.

Also,

$$v = v_p - \frac{P}{4\mu}\left(r^2 + 2r_p r - \frac{8}{3}\sqrt{r_p r^3} - \frac{1}{3}r_p^2\right).$$ (5.52)

The velocity profile of the flow is shown in Fig. 5.5.

Fig. 5.5: The velocity profile of Casson flow in a long circular cylindrical pipe.

To determine the flux of the flow,

$$
\begin{aligned}
Q &= \pi r_p^2 v_p + \int_{r_p}^{R} 2\pi r v \, dr \\[2mm]
&= \pi r_p^2 v_p + 2\pi \int_{r_p}^{R} \left[v_p r - \frac{P}{4\mu} \left(r^3 + 2 r_p r^2 - \frac{8}{3}\sqrt{r_p r^5} - \frac{r_p^2 r}{3} \right) \right] dr \\[2mm]
&= \pi r_p^2 v_p + \pi v_p (R^2 - r_p^2) - \frac{P\pi}{\mu} \left[\frac{1}{8}(R^4 - r_p^4) \right. \\[2mm]
&\quad \left. + \frac{r_p}{3}(R^3 - r_p^3) - \frac{r_p^2}{12}(R^2 - r_p^2) - \frac{8}{21}\sqrt{r_p}(R^{\frac{7}{2}} - r_p^{\frac{7}{2}}) \right] \\[2mm]
&= \frac{P\pi}{\mu} R^4 \left[\left(\frac{1}{4} + \frac{1}{2}\frac{r_p}{R} - \frac{2}{3}\sqrt{\frac{r_p}{R}} - \frac{1}{12}\frac{r_p^2}{R^2} \right) - \frac{1}{8}\left(1 - \frac{r_p^4}{R^4}\right) \right. \\[2mm]
&\quad \left. - \frac{1}{3}\frac{r_p}{R}\left(1 - \frac{r_p^3}{R^3}\right) + \frac{1}{12}\frac{r_p^2}{R^2}\left(1 - \frac{r_p^2}{R^2}\right) + \frac{8}{21}\sqrt{\frac{r_p}{R}}\left(1 - \sqrt{\frac{r_p^7}{R^7}}\right) \right] \\[2mm]
&= \frac{\pi P}{8\mu} R^4 \left(1 - \frac{16}{7}c_p^{\frac{1}{2}} + \frac{4}{3}c_p - \frac{1}{21}c_p^4 \right)
\end{aligned}
\tag{5.53}
$$

where, as before

$$
c_p = \frac{r_p}{R}.
\tag{5.54}
$$

Equation (5.53) can be written as

$$
Q = \frac{\pi P R^4}{8\mu} f(c_p),
\tag{5.55}
$$

where

$$\frac{Q}{Q_0} = f(c_p) = 1 - \frac{16}{7}c_p^{\frac{1}{2}} + \frac{4}{3}c_p - \frac{1}{21}c_p^4 \tag{5.56}$$

and Q_0 denotes the flux when there is no plug flow. The variation of $f(c_p)$ with c_p is shown in Fig. 5.6. It is seen that the flux of the flow decreases rapidly with increasing c_p. For $c_p > 1$, there is no flow.

Fig. 5.6: Variation of $f(c_p)$ with c_p.

Note that when there is no yield stress, that is, $\tau_0 = 0$, the Casson equation reduces to

$$\tau = \mu\dot{\gamma} \tag{5.57}$$

which is the Newtonian case. Putting $\tau_0 = 0$, i.e., $c_p = 0$ into Eq. (5.55) gives

$$Q = \frac{P\pi}{8\mu}R^4 \tag{5.58}$$

which coincides with the Poiseuille's equation, Eq. (4.13). Thus, Poiseuille's model may be regarded as a particular case of Casson's model.

5.6.4 *Further Analysis of the Casson Model*

In his paper, Scott Blair [99] used the results from an experiment carried out by F.A. Glover on human blood and plotted a graph of square-root of strain rate against square-root of shear stress. The curve shows remarkable linearity, with a non-zero value for the intercept on the stress axis (see Fig. 5.7). In the same paper, Scott Blair also cited the works of Kümin, Suter, Bircher and Müller, who presented similar results for animal blood (see Fig. 5.8).

Fig. 5.7: Scott Blair's graph for healthy human blood.

This linearity was further supported by the results of Merrill *et al.* [70]. Merrill used two different methods to obtain data for suspensions of red cells in plasma of two different hematocrits. All four sets of data produced graphs which are linear and have a positive stress intercept (see Fig. 5.9).

Another reason for the use of the Casson model is that it not only agrees with experimental results, but agrees over a large range of shear rates as well. Charm and Kurland [17] demonstrated the application of Casson's equation over the range of 2 to $100,000 \text{ sec}^{-1}$ using a series of viscometers. The results indicated that using Casson's equation, it was possible to extrapolate blood viscometry information obtained at shear ranges of 5 to 200 sec^{-1} to shear rates of 10,000 to $100,000 \text{ sec}^{-1}$ with less than 5 per cent error.

5.7 Flow of Non-Newtonian Fluids in Elastic Tubes

We will now consider the effect of the elasticity of the tube wall while considering flow of non-Newtonian fluids in tubes. If the wall is elastic then the fluid velocity vector will have both axial and radial component. Further, the pressure gradient can also have both axial and radial components. Thus, it is important that we incorporate the elasticity of the tube wall into the models of non-Newtonian fluids discussed in Sec. 5.6.

Fig. 5.8: Two samples of cow's blood (Kümin).

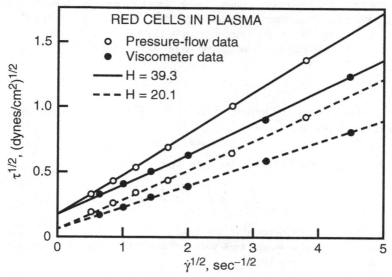

Fig. 5.9: Pressure-flow data and viscometer data on square-root coordinates.

5.7.1 *Power-Law Model Using Linear Elastic Theory*

Consider an elastic tube of length L with radius a which varies with changes in transmural pressure difference. Let the fluid enter the tube with inlet pressure p_1 and leave it with outlet pressure p_2, and the pressure outside be p_0 (see Fig. 5.10). We denote the distance along the tube from inlet end, A, by x.

Fig. 5.10: Fluid flow in an elastic tube.

The pressure inside the tube, $p(x)$, decreases from $p(0) = p_1$, to $p(L) = p_2$. The elastic tube will expand or contract at x, depending on the pressure difference $p(x) - p_0$, known as transmural pressure difference. As such, the radius of the tube is not a constant and hence, the cross-section of the tube may deform.

It is assumed that the flux Q is related to pressure gradient $\frac{dp}{dx}$ by the relation

$$Q = -\sigma \frac{dp}{dx} \tag{5.59}$$

where σ, known as the conductivity of the tube, will be a function of the pressure difference, $p(x) - p_0$, i.e.,

$$\sigma = \sigma[p(x) - p_0]. \tag{5.60}$$

Integrating Eq. (5.59) with respect to x, we have

$$Q = \frac{1}{L} \int_{p_2 - p_0}^{p_1 - p_0} \sigma(p') \, dp' \tag{5.61}$$

where, as before, we have denoted $p' = p - p_0$.

In the special case of a circular tube with radius a, a function of p', the conductivity for Newtonian fluid is given by

$$\sigma(p') = \frac{\pi}{8\mu} a^4(p'). \tag{5.62}$$

Thus, from Eq. (5.61) we obtain

$$Q = \frac{\pi}{8\mu L} \int_{p_2-p_0}^{p_1-p_0} a^4(p') \, dp' \qquad (5.63)$$

which, of course, reduces to Poiseuille's formula when the radius a is a constant.

Equation (5.63) can be solved if we know the form of the function $a(p')$. If the hoop stress or tension $T(a)$ in the tube wall is known as a function of a, then $a(p')$ can be found using the equilibrium condition

$$p(x) - p_0 = \frac{T}{a}. \qquad (5.64)$$

At this point it is necessary to know how the radius of a tube varies with pressure. Roach and Burton [86] determined the static pressure volume relation of a 4 cm long piece of human external iliac artery, and converted it into a tension versus length curve (see Fig. 5.11).

Fig. 5.11: The elastic diagram of a human iliac artery using percentage elongation.

Then, Rubinow and Keller [90] used the least squares method to fit this curve with the equation

$$T(a) = \left[t_1 \left(\frac{a}{a_0} - 1 \right) + t_2 \left(\frac{a}{a_0} - 1 \right)^5 \right] \qquad (5.65)$$

where T = tension in arterial wall
$\quad a$ = radius of artery
$\quad a_0 = 0.216$ cm (unstretched radius of artery)
$\quad t_1 = 13.0$ g/cm
$\quad t_2 = 300$ g/cm.

Using the relation in Eq. (5.65), Eq. (5.64) becomes

$$p' = p(x) - p_0 = \frac{1}{a}\left[t_1\left(\frac{a}{a_0} - 1\right) + t_2\left(\frac{a}{a_0} - 1\right)^5\right]. \tag{5.66}$$

Differentiation gives

$$\frac{dp'}{da} = \frac{t_1}{a^2} + \frac{t_2}{a^2}\left(\frac{a}{a_0} - 1\right)^4\left(\frac{4a}{a_0} + 1\right). \tag{5.67}$$

Therefore, with the use of Eq. (5.67), the definite integral for flux Q appearing in Eq. (5.63) can be evaluated to give

$$Q = \frac{\pi a_0^3}{8\mu L}[g(\alpha_1) - g(\alpha_2)] \tag{5.68}$$

where

$$g(\alpha) = \frac{1}{3}t_1\alpha^3 + t_2\left(\frac{1}{2}\alpha^8 - \frac{15}{7}\alpha^7 + \frac{10}{3}\alpha^6 - 2\alpha^5 + \frac{1}{3}\alpha^3\right) \tag{5.69}$$

and α_1, α_2 are non-dimensional parameters, given by

$$\alpha_1 = \frac{a_1}{a_0}, \qquad \alpha_2 = \frac{a_2}{a_0} \tag{5.70}$$

where a_1 and a_2 are values of a, corresponding to $(p_1 - p_0)$ and $(p_2 - p_0)$ respectively. Figure 5.12 shows the variation of $g(\alpha)$ with respect to α.

The above analysis refers to a Newtonian viscous fluid. We now consider the fluid in the tube obeying the power-law given by Eq. (5.13). We have the relation between the flux and the pressure gradient, given by Eq. (5.59). Comparing this with Eq. (5.28), we conclude

$$\sigma(p') = \frac{n\pi}{\eta(3n+1)}a^{3+\frac{1}{n}}, \quad 0 < n < 1 \tag{5.71}$$

where

$$\eta = (2\mu)^{\frac{1}{n}}. \tag{5.72}$$

Thus, using Eqs. (5.61) and (5.71), we obtain

$$Q = \frac{n\pi}{\eta(3n+1)L}\int_{p_2-p_0}^{p_1-p_0} a^{\frac{3n+1}{n}}\,dp'. \tag{5.73}$$

Let us denote

$$K = \frac{n\pi}{\eta(3n+1)L}. \tag{5.74}$$

Assuming as before the tension-strain relation of the tube given by Eq. (5.65), and proceeding as before, we finally obtain

$$Q = Ka_0^s[g(\alpha_1) - g(\alpha_2)] \tag{5.75}$$

Fig. 5.12: Plot of $g(\alpha)$ vs α.

where

$$g(\alpha) = \frac{t_1 \alpha^s}{s} + t_2 \left[\frac{4\alpha^{s+5}}{s+5} - \frac{15\alpha^{s+4}}{s+4} + \frac{20\alpha^{s+3}}{s+3} - \frac{10\alpha^{s+2}}{s+2} + \frac{\alpha^s}{s} \right] \qquad (5.76)$$

and

$$s = \frac{2n+1}{n}, \qquad 0 < n < 1. \qquad (5.77)$$

At $n = 1$, Eq. (5.76) coincides with Eq. (5.69), the case for a Newtonian fluid. Since the model largely depends on the function $g(\alpha)$ for various values of the power index, n, the graph of $g(\alpha)$ for a number of values of n is shown in Fig. 5.13. It should be noted that the graph for $n = 1$ is exactly the same as that shown in Fig. 5.12.

Equation (5.75) is then the flux for the flow of a non-Newtonian fluid with variable viscosity obeying the power-law, in an elastic tube. This model may be applied to model the flow of blood in arteries. However, we need to know the value of n, the index of the power law, before this equation can be used. It should also be noted that in the equation, η is not the viscosity — η here is a constant, acting as a parameter when the "power" n is being decided. Hence, to apply the above model, prior knowledge of the viscosity of blood in the area considered is essential and this will in turn determine the appropriate values of n and μ.

The above analysis is based on the choice of a polynomial form as an approximating function for the tension-strain relationship as given by Eq. (5.65). Let us now consider the case when an exponential function is chosen as an approximating function

Fig. 5.13: Graph of $g(\alpha)$ for various values of n, power index.

to the experimental data. Suppose the tension $T(a)$ takes the form (Mazumdar *et al.* [69])

$$T(\alpha) = Ae^{k\alpha} + B, \tag{5.78}$$

where

$$\alpha = \frac{a}{a_0}$$

and k, A and B are constant parameters to be determined.

Because a_0 is the initial radius, we will have an initial condition:

$$T(\alpha) = 0 \quad \text{at} \quad \alpha = 1.$$

Therefore,

$$T(\alpha) = A(e^{k\alpha} - e^k) \tag{5.79}$$

so that A is positive. Also, from the plot of the experimental data points (Fig. 5.11), we expect k to be positive since the curve is monotonically increasing. Further, we can employ the method of least squares to find the values of A and k in Eq. (5.78). The values thus obtained are

$$A = 0.007435, \quad k = 5.262500. \tag{5.80}$$

Hence, with these values of A and k, Eq. (5.79) becomes

$$T(\alpha) = 0.007435 \left(e^{5.2625\frac{a}{a_0}} - e^{5.2625} \right). \tag{5.81}$$

Proceeding as before, the value of Q can be calculated from Eq. (5.73) as

$$Q = KA \int_{a_2}^{a_1} a^{\frac{n+1}{n}} \left[\left(\frac{ka}{a_0} - 1 \right) e^{\frac{ka}{a_0}} + e^k \right] da. \tag{5.82}$$

However, the above integral can be evaluated numerically, for instance, by using Simpson's Quadrature Formula, or the Trapezoidal Rule.

We thus obtained results using the polynomial form as well as exponential form of representing the tension-strain relationship of the vessel wall. Figure 5.14 illustrates how well the two results approximate the experimental data, given by Roach and Burton [86], and displayed in Table 5.1.

Fig. 5.14: Graph of tension against strain, experimental data, (X), polynomial approximating function, (P), and exponential approximating function, (E).

Table 5.1: Relationship between tension and strain.

T (g/cm)	4	8	13	20	22	28	33	40
a/a_0	1.22	1.35	1.45	1.50	1.55	1.57	1.60	1.64
T (g/cm)	44	60	71	83	95	109	132	
a/a_0	1.67	1.71	1.74	1.77	1.80	1.83	1.86	

It is clear from this figure that the exponential function fits the experimental data better than the polynomial function. This is almost an expected result, since the exponential is in fact an infinite series, whereas the polynomial is merely a truncated part of the series, ignoring higher order terms.

5.7.2 *Casson Model Using Linear Elastic Theory*

Consider now the flow in an elastic tube of a fluid obeying Casson's law given by Eq. (5.40). We have the relation between the flux and the pressure gradient given by Eq. (5.59). Comparing this with Eq. (5.53), we conclude

$$\sigma(p') = \frac{\pi}{8\mu} a^4 \left(1 - \frac{16}{7} c_p^{\frac{1}{2}} + \frac{4}{3} c_p - \frac{1}{21} c_p^4 \right). \tag{5.83}$$

We thus obtain the value of the flux, Q for Eq. (5.61) as

$$Q = \frac{\pi}{8\mu L} \int_{p_2 - p_0}^{p_1 - p_0} a^4 \left(1 - \frac{16}{7} c_p^{\frac{1}{2}} + \frac{4}{3} c_p - \frac{1}{21} c_p^4 \right) dp' \tag{5.84}$$

or, using the relation in Eq. (5.54), we get

$$Q = \frac{\pi}{8\mu L} \int_{p_2 - p_0}^{p_1 - p_0} \left(a^4 - \frac{16}{7} \sqrt{r_p a^7} + \frac{4}{3} r_p a^3 - \frac{1}{21} r_p^4 \right) dp'. \tag{5.85}$$

Consider now the exponential form of tension-extension for the tube wall as given in Eq. (5.81). We have from Eq. (5.79)

$$p' = p(x) - p_0 = \frac{A}{a} \left(e^{\frac{ka}{a_0}} - e^k \right). \tag{5.86}$$

Hence,

$$\frac{dp'}{da} = A \left[e^{ca} \left(\frac{c}{a} - \frac{1}{a^2} \right) + \frac{e^k}{a^2} \right], \tag{5.87}$$

where

$$c = \frac{k}{a_0}. \tag{5.88}$$

Equation (5.85) now takes the form

$$
\begin{aligned}
Q &= \frac{A\pi}{8\mu L} \int_{a_2}^{a_1} \left(a^4 - \frac{16}{7} \sqrt{r_p a^7} + \frac{4}{3} r_p a^3 - \frac{1}{21} r_p^4 \right) \left(e^{ca} \left(\frac{c}{a} - \frac{1}{a^2} \right) + \frac{e^k}{a^2} \right) da \\
&= \frac{A\pi}{\mu L} \int_{a_2}^{a_1} \left[e^k \left(\frac{a^2}{8} + \frac{r_p}{6} a - \frac{2}{7} \sqrt{r_p a^3} - \frac{r_p^4}{168} \frac{1}{a^2} \right) + e^{ca} \left(\frac{c}{8} a^3 + \frac{r_p c}{6} a^2 \right. \right. \\
&\qquad \left. \left. - \frac{2c}{7} \sqrt{r_p a^5} - \frac{r_p^4 c}{168} \frac{1}{a} - \frac{a^2}{8} - \frac{r_p}{6} a + \frac{2}{7} \sqrt{r_p a^3} + \frac{r_p^4}{168} \frac{1}{a^2} \right) \right] da
\end{aligned}
\tag{5.89}
$$

where $A = 0.007435$
$k = 5.2625$
$a_0 = 0.216$ cm
$c = \frac{k}{a_0}$.

The above integral, after numerical evaluation, gives the value of flux for Casson's model in an elastic tube. The results in this chapter give complete information on the laminar flow in circular cylindrical tubes.

Chapter 6

MODELS FOR OTHER FLOWS

6.1 Introduction

In this chapter some applications of the basic equations derived earlier are demonstrated and a number of models of other flows are discussed. In particular, a model of oxygen diffusion from blood vessel to tissue, as proposed by Krogh [55], a model for fluid movement in renal tubules in kidneys, and diffusion in artificial kidneys, a model for the measurement of the extravascular space by the indicator dilution method, and finally a model for the peristaltic flow, which is the motion of a fluid in an elastic tube when a wave of contraction and expansion travels along the wall of the tube, are discussed in detail.

A comprehensive study of the dynamics of flows of other biofluids is beyond the scope of this book. There are many books and papers on this subject, amongst which the following are worth mentioning; Fung [28,29,30], Kapur [50], Lighthill [60,61], and Nevaril et al. [77]. We will emphasise in this chapter a quantitative way of assessing certain aspects of these models.

One of the basic mechanisms in biological systems is that of the process of diffusion in the human system. Accordingly, in the next section we discuss this process in order to explain the diffusion of oxygen from blood in capillaries to the surrounding tissues.

6.2 The Krogh Model of Oxygen Diffusion from Blood Vessel to Tissue

In this model, the capillary blood vessel is represented by a straight circular cylindrical tube and the tissue is represented by a stationary concentric cylindrical tube (see Fig. 6.1). A large number of identical tubes are closely packed to represent the tissue of an organ (Fig. 6.2).

The model is used to study the transport of molecular oxygen from the blood plasma to the skeletal muscle tissues or lungs or brain, across the capillary walls. The model will give an indication of the effect of the velocity of blood flow in capillaries, axial and radial diffusion of oxygen in blood, oxygen diffusivity in tissues, and permeability of capillary walls.

Fig. 6.1: Krogh tissue cylinder.

Fig. 6.2: Closely packed identical tubes.

The model hypothesizes that, in a given portion of tissue, all capillaries and surrounding tissues are of equal diameter and are homogeneously dispersed in the tissue. In the capillary vessel, oxygen transport takes place both by convection and by diffusion. In the tissue there is diffusion of oxygen, as well as consumption of oxygen by the tissue cells.

Since diffusion is the main mechanism, we will first introduce the basic equations of the process of diffusion. For more detail see Mazumdar [68]. Let $c(\mathbf{x}, t)$ be the concentration of a solute or the amount of solute per unit volume at the point $\mathbf{x}$ at time t. Due to the concentration gradient, there is a flow of solute given by the flux vector or the current density vector $\mathbf{J}$. This vector consists of two parts: diffusive $\mathbf{J}_{diff}$ and convective $\mathbf{J}_{conv}$. The former obeys Fick's law, according to which the flux is proportional to the concentration gradient. The latter is proportional to the product of velocity and concentration. Hence, we have

$$\mathbf{J}_{diff} = -D\nabla c, \tag{6.1}$$

$$\mathbf{J}_{conv} = c\mathbf{q} \tag{6.2}$$

where $\mathbf{q}$ (with components u, v, w) is the flow velocity vector and D is the coefficient of diffusion. Combining these two current densities we have the total flux vector $\mathbf{J} = (J_x, J_y, J_z)$ given by

$$J_x = -D_x\frac{\partial c}{\partial x} + uc, \quad J_y = -D_y\frac{\partial c}{\partial y} + vc, \quad J_z = -D_z\frac{\partial c}{\partial z} + wc, \tag{6.3}$$

where D_x, D_y, D_z are the coefficients of diffusion in the x, y, z directions respectively.

We will now apply the conservation principle to the transport of solute. The conservation principle is a fundamental principle of nature and all natural phenomena must, whether it is physical, biological or chemical, conform with this principle.

Consider a very small cubical volume of solution (as shown in Fig. 6.3).

Fig. 6.3: Conservation of solute transport into and out of a small volume element.

Conservation of solute transport into and out of this volume element states that, within a given infinitesimal volume element of a solution in which solute currents exist, whatever the cause of the currents may be, the rate at which matter accumulates or disappears within the region is equal to the net flux across the surface bounding the infinitesimal region. This is stated by the following expression:

$$\left\{ J_x dydz - \left(J_x + \frac{\partial J_x}{\partial x}dx \right)dydx \right\} + \left\{ J_y dxdz - \left(J_y + \frac{\partial J_y}{\partial y}dy \right)dxdz \right\}$$

$$+ \left\{ J_z dxdy - \left(J_z + \frac{\partial J_z}{\partial z}dz \right)dxdy \right\} = \left(g + \frac{\partial c}{\partial t} \right)dxdydz \tag{6.4}$$

where the L.H.S. of the above equation is the net diffusive flux of solute across the surface bounding the volume element and the R.H.S. is the increment per unit time of the concentration of the infinitesimal volume element. Here, we assumed that due to chemical reaction in the volume element, the metabolite is consumed at a rate g moles per unit volume per unit time. By simplifying Eq. (6.4) we obtain

$$g + \frac{\partial c}{\partial t} + \frac{\partial J_x}{\partial x} + \frac{\partial J_y}{\partial y} + \frac{\partial J_z}{\partial z} = 0. \tag{6.5}$$

Substituting Eqs. (6.3) into Eq. (6.5), we obtain

$$g + \frac{\partial c}{\partial t} = \frac{\partial}{\partial x}\left(D_x \frac{\partial c}{\partial x}\right) + \frac{\partial}{\partial y}\left(D_y \frac{\partial c}{\partial y}\right) + \frac{\partial}{\partial z}\left(D_z \frac{\partial c}{\partial z}\right)$$
$$-c\left(\frac{\partial u}{\partial x} + \frac{\partial v}{\partial y} + \frac{\partial w}{\partial z}\right) - u\frac{\partial c}{\partial x} - v\frac{\partial c}{\partial y} - w\frac{\partial c}{\partial z} \tag{6.6}$$

which is the basic equation for diffusion. If the fluid is incompressible, then

$$\frac{\partial u}{\partial x} + \frac{\partial v}{\partial y} + \frac{\partial w}{\partial z} = 0. \tag{6.7}$$

If, in addition, the coefficient of diffusion D is a constant, then Eq. (6.6) is reduced to

$$\frac{Dc}{Dt} \equiv \frac{\partial c}{\partial t} + u\frac{\partial c}{\partial x} + v\frac{\partial c}{\partial y} + w\frac{\partial c}{\partial z}$$
$$= D\left(\frac{\partial^2 c}{\partial x^2} + \frac{\partial^2 c}{\partial y^2} + \frac{\partial^2 c}{\partial z^2}\right) - g. \tag{6.8}$$

We will now discuss this equation in the capillary region as well as in the tissue region.

6.2.1 *Capillary Blood Vessel Region*

In the capillaries, assume the distribution of oxygen is axisymmetric. Let $v(r,t)$ denote the velocity of blood in the fully developed flow, $c(r,z,t)$ be the concentration of oxygen, $d(c)$ be the rate of generation of oxygen per unit volume and r_c be the radius of a capillary (see Fig. 6.1). Hence, after using cylindrical polar coordinates, Eq. (6.8) gives

$$\frac{Dc}{Dt} = D_b\left(\frac{\partial^2 c}{\partial r^2} + \frac{1}{r}\frac{\partial c}{\partial r} + \frac{\partial^2 c}{\partial z^2}\right) + d(c), \qquad 0 < r < r_c \tag{6.9}$$

where D_b denotes the overall coefficient of diffusion of oxygen in the blood.

For large capillaries, Eq. (6.9) is satisfactory, but for small capillaries, it is usually preferable to use two diffusivity coefficients, i.e., D_r in the radial direction and D_b in the axial direction.

6.2.2 *Tissue Region*

In this region, let D_t denote the coefficient of diffusion for oxygen. The velocity of motion of the tissue is zero and hence, there is no convective term. In this region, the basic diffusion equation (6.8) becomes

$$\frac{\partial c}{\partial t} = D_t \left(\frac{\partial^2 c}{\partial r^2} + \frac{1}{r}\frac{\partial c}{\partial r} + \frac{\partial^2 c}{\partial z^2} \right) - g(c), \qquad r_c \le r \le r_t. \tag{6.10}$$

It is assumed that the oxygen consumption rate $g(c)$ appearing in the above equation follows Michaelis-Menton kinetics [71] so that

$$g(c) = \frac{\lambda c}{\mu + c} \tag{6.11}$$

where λ and μ are constants.

If c is large in Eq. (6.11), $g(c)$ becomes a constant, say g_0, whereas if c is small, $g(c)$ will be of the form $\lambda_1 c$, where λ_1 is a constant.

6.2.3 *Boundary Conditions*

The boundary conditions for this situation are as follows:

(i) Due to the symmetry of the capillary flow about the axis, the radial component of the concentration gradient vanishes, i.e.,

$$\frac{\partial c}{\partial r} = 0 \quad \text{at} \quad r = 0. \tag{6.12a}$$

(ii) There is no flux at the surface of the tissue, i.e.,

$$\frac{\partial c}{\partial r} = 0 \quad \text{at} \quad r = r_t. \tag{6.12b}$$

(iii) The continuity condition at the interface requires that the flux of oxygen must be continuous at the blood-tissue interface, so that

$$\left(D_b \frac{\partial c}{\partial r} \right)_{\text{blood}} = \left(D_t \frac{\partial c}{\partial r} \right)_{\text{tissue}} \qquad \text{at} \quad r = r_c. \tag{6.12c}$$

6.2.4 *Krogh's Steady-State Model*

Krogh [55] obtained the steady-state solution for the tissue region, assuming that $g(c)$ in Eq. (6.10) is a constant g_0 and there is no axial diffusion.

Under these assumptions, Eq. (6.10) reduces, in steady state, to

$$\frac{\partial^2 c}{\partial r^2} + \frac{1}{r}\frac{\partial c}{\partial r} = \frac{g_0}{D_t}. \tag{6.13}$$

The solution to this equation, subjected to the boundary condition (6.12b), together with the condition that $c = c_0$ at $r = r_c$, is

$$c = c_0 + \frac{g_0}{D_t}\left(\frac{r^2 - r_c^2}{4} - \frac{r_t^2}{2}\ln\frac{r}{r_c}\right). \tag{6.14}$$

The above equation can be written as

$$\frac{c_0 - c}{g_0 r_c^2 D_t^{-1}} = \frac{1}{2}R^2\ln x - \frac{1}{4}(x^2 - 1) \tag{6.15}$$

where

$$R = \frac{r_t}{r_c}, \quad x = \frac{r}{r_c}, \quad 1 \le x \le R. \tag{6.16}$$

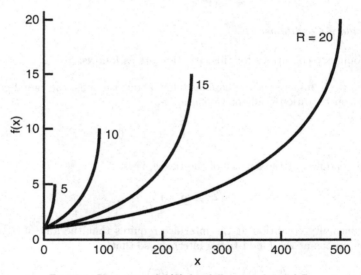

Fig. 6.4: Variation of $f(x)$ for different values of R.

Denoting

$$\frac{1}{2}R^2\ln x - \frac{1}{4}(x^2 - 1) = f(x) \tag{6.17}$$

the variation of $f(x)$ with x for various values of R is shown in Fig. 6.4.

6.2.5 *Blum's Steady-State Model*

Blum [12] relaxed some of the conditions imposed in Krogh's model and obtained the steady-state solution for the tissue region. The basic equations and the boundary condition for this model are as follows:

$$D_t \frac{1}{r} \frac{\partial}{\partial r} \left(r \frac{\partial c}{\partial r} \right) = kc, \quad r_c \leq r \leq r_t, \tag{6.18}$$

$$\frac{\partial c}{\partial r} = 0 \quad \text{at} \quad r = r_t, \tag{6.19}$$

$$\bar{v} \frac{\partial \bar{c}}{\partial z} = \frac{2}{r_c} \left(D_t \frac{\partial c}{\partial r} \right)_{r=r_c}, \tag{6.20}$$

$$\left(D_t \frac{\partial c}{\partial r} \right)_{r=r_c} = \left(D_b \frac{\partial c}{\partial r} \right)_{r=r_c} = P(c_0 - \bar{c}), \tag{6.21}$$

$$\bar{c} = c_{in} \quad \text{at} \quad z = 0, \quad 0 \leq r \leq r_c, \tag{6.22}$$

where $\bar{v}$ is the average velocity of blood in the capillary, c_{in} is the concentration of oxygen at the capillary inlet, and $\bar{c}(z)$ is the average concentration, given by

$$\bar{c}(z) = \frac{1}{\pi r_c^2} \int_0^{r_c} c(r, z) \, 2\pi r \, dr. \tag{6.23}$$

Clearly, Blum's model considered first-order kinetics when $g(c) = kc$, and considered the axial variation of oxygen concentration in the capillary blood vessel.

We notice that the basic equation (6.18) is the modified Bessel equation

$$\frac{\partial^2 c}{\partial r^2} + \frac{1}{r} \frac{\partial c}{\partial r} - \frac{k}{D_t} c = 0 \tag{6.24}$$

which reduces to

$$\frac{\partial^2 c}{\partial r^2} + \frac{1}{r} \frac{\partial c}{\partial r} - \frac{\lambda^2}{r_c^2} c = 0 \tag{6.25}$$

if we denote

$$\lambda^2 = \frac{k r_c^2}{D_t}. \tag{6.26}$$

The solution of Eq. (6.25) is given by

$$c(r, z) = A(z) I_0 \left(\frac{\lambda r}{r_c} \right) + B(z) K_0 \left(\frac{\lambda r}{r_c} \right) \tag{6.27}$$

where I_0 and K_0 are the modified Bessel functions of zero order, and $A(z), B(z)$ are arbitrary functions of z.

Using the boundary condition (6.19), we get

$$A(z)I_1\left(\frac{\lambda r_t}{r_c}\right) - B(z)K_1\left(\frac{\lambda r_t}{r_c}\right) = 0. \tag{6.28}$$

Using this result in Eq. (6.27) we obtain

$$c(r,z) = A(z)\left[I_0\left(\frac{\lambda r}{r_c}\right) + \frac{I_1\left(\frac{\lambda r_t}{r_c}\right)}{K_1\left(\frac{\lambda r_t}{r_c}\right)}K_0\left(\frac{\lambda r}{r_c}\right)\right]. \tag{6.29}$$

If we put $r = r_c$, we obtain

$$c(r_c,z) = A(z)\left[I_0(\lambda) + \frac{I_1\left(\frac{\lambda r_t}{r_c}\right)}{K_1\left(\frac{\lambda r_t}{r_c}\right)}K_0(\lambda)\right] = c_0(z) \quad (\text{say}). \tag{6.30}$$

So substituting the value of $A(z)$ obtained from Eq. (6.30) into Eq. (6.29), we get

$$c(r,z) = c_0(z)F\left(\frac{r}{r_c}, \frac{r_t}{r_c}, \lambda\right) \tag{6.31}$$

where

$$F\left(\frac{r}{r_c}, \frac{r_t}{r_c}, \lambda\right) = \frac{I_0\left(\frac{\lambda r}{r_c}\right)K_1\left(\frac{\lambda r_t}{r_c}\right) + I_1\left(\frac{\lambda r_t}{r_c}\right)K_0\left(\frac{\lambda r}{r_c}\right)}{I_0(\lambda)K_1\left(\frac{\lambda r_t}{r_c}\right) + I_1\left(\frac{\lambda r_t}{r_c}\right)K_0(\lambda)} \tag{6.32}$$

and

$$F\left(\frac{r}{r_c}, \frac{r_t}{r_c}, \lambda\right)_{r=r_c} = 1. \tag{6.33}$$

Differentiating Eq. (6.31) and putting $r = r_c$ we obtain

$$\left(\frac{\partial c}{\partial r}\right)_{r=r_c} = c_0(z)\frac{\lambda}{r_c}\left[\frac{I_1(\lambda)K_1\left(\frac{\lambda r_t}{r_c}\right) - I_1\left(\frac{\lambda r_t}{r_c}\right)K_1(\lambda)}{I_0(\lambda)K_1\left(\frac{\lambda r_t}{r_c}\right) + I_1\left(\frac{\lambda r_t}{r_c}\right)K_0(\lambda)}\right] \tag{6.34}$$

$$= -c_0\left(\frac{\lambda}{r_c}\right)B \quad (\text{say}) \tag{6.35}$$

where the quantity in Bessel functions in Eq. (6.34) is denoted by B.

Using Blum's boundary condition (6.21) we get

$$D_t c_0\left(\frac{\lambda}{r_c}\right)B + P(c_0 - \bar{c}) = 0 \tag{6.36}$$

or

$$\bar{c}(z) = c_0(z)(1 + \beta B) \tag{6.37}$$

where

$$\beta = \frac{\lambda D_t}{Pr_c} = \frac{(kD_t)^{\frac{1}{2}}}{P}. \tag{6.38}$$

Thus, from Eq. (6.31) we finally obtain Blum's steady-state solution in the tissue region

$$c(r, z) = \frac{\bar{c}(z)}{1 + \beta B} F\left(\frac{r}{r_c}, \frac{r_t}{r_c}, \lambda\right). \tag{6.39}$$

Clearly, when $P \to \infty$, the above solution gives

$$c_0(z) = \bar{c}. \tag{6.40}$$

6.3 Fluid Flow in Kidneys

6.3.1 *Introduction*

One of the main functions of kidneys is to maintain the chemical balance of blood by excreting the waste products such as urea, creatine and uric acid in the blood stream. If the urea is not removed, a disease known as *uremia* occurs which may sometimes be fatal. Sometimes, the impure blood is taken out of the body. The urea is then removed from it with the help of an external device known as a *haemodialyser*, and the purified blood is returned to the body. This process is known as the *dialysis of blood.*

An important diffusion process takes place in renal tubules in kidneys, in the form of diffusion of urea through the semi-permeable membrane of a dialyser. Considering the variation of urea concentration in the dialyzate, the problem reduces to solving two partial differential equations for the blood and the dialyzate regions.

Also, an important fluid flow takes place in the functional unit of the kidney known as the *nephron* or *renal tubule*. Each kidney has about a million of these tubules. Here, because of reabsorption of water and other low molecular weight substances along the walls of the renal tube we have both radial and axial velocity of flow.

6.3.2 *Diffusion Process in the Haemodialyser*

A haemodialyser, sometimes called an *artificial kidney*, is regarded as a concentric circular duct (Fig. 6.5). Blood flows on the inside, and outside, the fluid called dialyzate which is a solution of some chemicals in water, flows. The wall of the inner cylinder is a semi-permeable membrane that allows urea to diffuse through it. During the process of flow in the duct, blood loses urea which permeates through the duct wall into the dialyzate. By maintaining a continuous supply of fresh dialyzate, the concentration of urea in the dialyzate is maintained lower than that of the blood.

Fig. 6.5: Circular-duct haemodialyzer.

The basic governing equation is the diffusion equation given by Eq. (6.8) in cylindrical coordinates

$$D \left(\frac{\partial^2 c}{\partial r^2} + \frac{1}{r} \frac{\partial c}{\partial r} + \frac{\partial^2 c}{\partial z^2} \right) = \frac{\partial c}{\partial t} + v \frac{\partial c}{\partial z}, \tag{6.41}$$

where $c(r, z, t)$ is the concentration of urea in the blood and $v(r, t)$ is the velocity in the fully-developed flow. If we assume steady-state laminar flow in a straight duct and neglect the longitudinal diffusion term, we then have

$$D \left(\frac{\partial^2 c}{\partial r^2} + \frac{1}{r} \frac{\partial c}{\partial r} \right) = v_m \left(1 - \frac{r^2}{R^2} \right) \frac{\partial c}{\partial z}, \tag{6.42}$$

where v_m is the maximum velocity in the duct and R is the radius of the duct.

The boundary conditions for this flow are as follows:

(i) $\dfrac{\partial c}{\partial r} = 0$ at $r = 0,$ $\tag{6.43a}$

(ii) $c = c_{in}$ at $z = 0, \quad 0 \leq r \leq R,$ $\tag{6.43b}$

(iii) $-D \dfrac{\partial c}{\partial r} = P(c - c_d)$ at $r = R,$ $\tag{6.43c}$

where the last condition follows on the assumption of constant permeability P and constant concentration c_d in the dialyzate.

Introducing non-dimensional variables

$$\bar{c} = \frac{c - c_d}{c_{in} - c_d}, \quad \bar{r} = \frac{r}{R}, \quad P_e = \frac{v_m R}{D}, \quad \bar{z} = \frac{z}{R P_e} \tag{6.44}$$

where P_e is the dimensionless Peclet number, Eqs. (6.42) and (6.43) finally reduce to

$$\frac{\partial^2 \bar{c}}{\partial \bar{r}^2} + \frac{1}{\bar{r}} \frac{\partial \bar{c}}{\partial \bar{r}} = (1 - \bar{r}^2) \frac{\partial \bar{c}}{\partial \bar{z}}, \tag{6.45}$$

$$\frac{\partial \bar{c}}{\partial \bar{r}} = 0 \quad \text{at} \quad \bar{r} = 0, \tag{6.46a}$$

$$\bar{c} = 1 \quad \text{at} \quad \bar{z} = 0, \quad 0 \leq \bar{r} \leq 1, \tag{6.46b}$$

$$\frac{\partial \bar{c}}{\partial \bar{r}} + Sh_w \bar{c} = 0 \quad \text{at} \quad \bar{r} = 1, \quad \bar{z} > 0, \tag{6.46c}$$

where Sh_w is called the *Sherwood wall number*, given by

$$Sh_w = \frac{PR}{D}. \tag{6.47}$$

The solution to Eq. (6.45) with the boundary conditions (6.46) can be obtained in a variety of ways. If we use the method of separation of variables, assume the solution is of the form (Kapur [50])

$$\bar{c}(\bar{r}, \bar{z}) = \sum_{n=0}^{\infty} A_n R_n(\bar{r}) \exp\left(-\lambda_n^2 \bar{z}\right) \tag{6.48}$$

where the λ_n's are the eigenvalues, R_n's are eigenfunctions and coefficient A_n's are to be determined using the orthogonality properties of eigenfunctions.

Substituting Eq. (6.48) into Eq. (6.45) yields

$$\frac{d^2 R_n}{d\bar{r}^2} + \frac{1}{\bar{r}} \frac{dR_n}{d\bar{r}} + \lambda_n^2 (1 - \bar{r}^2) R_n = 0 \tag{6.49}$$

with boundary conditions

$$\frac{dR_n}{d\bar{r}} = 0 \quad \text{at} \quad \bar{r} = 0, \tag{6.50a}$$

$$\sum_{n=0}^{\infty} A_n R_n = 1 \quad \text{for} \quad 0 \leq \bar{r} \leq 1, \tag{6.50b}$$

$$\frac{dR_n}{d\bar{r}} + Sh_w R_n = 0 \quad \text{at} \quad \bar{r} = 1. \tag{6.50c}$$

The solution $R_n(\bar{r})$ to Eq. (6.49) satisfying the boundary conditions in Eqs. (6.50a–c) can thus be obtained in the form of a power series in $\bar{r}$, which when substituted into Eq. (6.48) gives the required concentration of urea in the blood in a haemodialyser. The same analysis can be made if instead of a circular-duct dialyser, we consider a parallel-plate dialyser.

6.3.3 *Flow in the Renal Tubule*

The renal tubule or nephron is the functional unit of the kidney. Each kidney has about one million of these tubules. The main function of these tubules is to filter the blood coming from the renal artery. During the course of filtration this filtrate is subsequently reabsorbed by the collecting ducts.

Due to this reabsorption, both radial and axial components of the velocity of flow exist (Fig. 6.6). Hence, if we neglect the inertial terms since the Reynolds number in this case is very much less than one (about 10^{-3}), we have the following basic equations corresponding to Eqs. (1.44)–(1.49) discussed in Chap. 1.

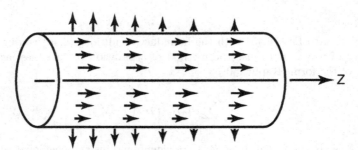

Fig. 6.6: Flow in a renal tubule.

$$\frac{1}{r}\frac{\partial}{\partial r}(rv_r) + \frac{\partial v_z}{\partial z} = 0, \tag{6.51}$$

$$\frac{\partial^2 v_r}{\partial r^2} + \frac{1}{r}\frac{\partial v_r}{\partial r} - \frac{v_r}{r^2} + \frac{\partial^2 v_r}{\partial z^2} = \frac{1}{\mu}\frac{\partial p}{\partial r}, \tag{6.52}$$

$$\frac{\partial^2 v_z}{\partial r^2} + \frac{1}{r}\frac{\partial v_z}{\partial r} + \frac{\partial^2 v_z}{\partial z^2} = \frac{1}{\mu}\frac{\partial p}{\partial z}. \tag{6.53}$$

The boundary conditions for this flow are as follows:

(i) $\dfrac{\partial v_z}{\partial r} = 0, \quad v_r = 0 \quad \text{at} \quad r = 0,$ \hfill (6.54a)

(ii) $v_z = 0, \quad v_r = \phi(z) \quad \text{at} \quad r = R.$ \hfill (6.54b)

Eliminating p between Eqs. (6.52) and (6.53) and introducing the stream function ψ, such that

$$v_z = \frac{1}{r}\frac{\partial \psi}{\partial r}, \qquad v_r = -\frac{1}{r}\frac{\partial \psi}{\partial z}, \tag{6.55}$$

we finally obtain, as shown in Eq. (1.48)

$$\nabla_1^4 \psi = 0 \tag{6.56}$$

where

$$\nabla_1^2 \equiv \frac{\partial^2}{\partial r^2} - \frac{1}{r}\frac{\partial}{\partial r} + \frac{\partial^2}{\partial z^2}. \tag{6.57}$$

In order to solve Eq. (6.56), let us try a solution of the form

$$\psi(r, z) = f(r)\left(a_0 z + \frac{1}{2}a_1 z^2\right) + g(r). \tag{6.58}$$

The above form of the solution is based on the assumption that the function $\phi(z)$, which is the value of v_r at $r = R$, is a linear function in z.

We thus have, using Eq. (6.55)

$$v_r = -\frac{1}{r}f(r)(a_0 + a_1 z), \tag{6.59}$$

$$v_z = \frac{1}{r}f'(r)(a_0 z + \frac{1}{2}a_1 z^2) + g'(r) \tag{6.60}$$

and Eq. (6.56) gives

$$(a_0 z + \frac{1}{2}a_1 z^2)D^2 f(r) + D^2 g(r) + 2a_1 D f(r) = 0 \tag{6.61}$$

where the operator D is given by

$$D \equiv \frac{d^2}{dr^2} - \frac{1}{r}\frac{d}{dr}. \tag{6.62}$$

We thus have,

$$D^2 f(r) = 0, \tag{6.63}$$

$$D^2 g(r) + 2a_1 D f(r) = 0. \tag{6.64}$$

Solving Eqs. (6.63) and (6.64) using the required boundary conditions, we finally obtain

$$f(r) = \frac{2r^2}{R} - \frac{r^4}{R^3} \tag{6.65}$$

and

$$g(r) = \left(\frac{a_1 R}{12} - \frac{Q_0}{\pi R^2}\right)r^2 + \left(\frac{Q_0}{2\pi R^4} - \frac{a_1}{6R}\right)r^4 + \frac{a_1}{12R^3}r^6, \tag{6.66}$$

where Q_0 is the value of flux at the entrance ($z = 0$), i.e.,

$$Q_0 = \int_0^R 2\pi r v_z \, dr. \tag{6.67}$$

We can thus calculate the velocity components v_r and v_z, and the pressure $p(r, z)$. It can be shown that in this case

$$v_z = \frac{2Q(z)}{\pi R^2}\left[\left(1 - \frac{r^2}{R^2}\right) - \frac{a_1 R}{2}\left(\frac{1}{3} - \frac{r^2}{R^2}\right)\right] \tag{6.68}$$

in contrast to the expression for Poiseuille's flow

$$v_z = \frac{2Q}{\pi R^2}\left(1 - \frac{r^2}{R^2}\right)$$ (6.69)

and

$$p(r,z) = p_0 - \frac{4\mu}{R^3}(a_0 + a_1 z)r^2 - \mu\left(\frac{4a_1}{3R} + \frac{8\overline{Q}(z)}{\pi R^4}\right)z$$ (6.70)

where

$$Q(z) = Q_0 - \pi R(2a_0 z + a_1 z^2)$$ (6.71)

and

$$\overline{Q}(z) = \int_0^z Q(z)\,dz.$$ (6.72)

The above results are obtained assuming a particular form of the stream function given by Eq. (6.58). Let us now consider another form of the stream function, given by

$$\psi(r,z) = f(r)e^{-\alpha z} + g(r)$$ (6.73)

where we have assumed that the radial velocity at the wall of the cylinder decreases exponentially with z. Proceeding as before, we obtain

$$D_1^2 f(r) = 0,$$ (6.74)

$$D^2 g(r) = 0$$ (6.75)

where

$$D_1 \equiv \frac{d^2}{dr^2} - \frac{1}{r}\frac{d}{dr} + \alpha^2.$$ (6.76)

Solving Eqs. (6.74) and (6.75) we get the functions $f(r)$ and $g(r)$, then using Eq. (6.55) we get velocity components v_r and v_z, and the pressure $p(r,z)$. It can be shown that in this case

$$v_r = \frac{J_1(\alpha R)J_1(\alpha r) - \frac{r}{R}J_0(\alpha R)J_2(\alpha r)}{J_1^2(\alpha R) - J_0(\alpha R)J_2(\alpha R)}v_0 e^{-\alpha z},$$ (6.77)

$$v_z = \frac{J_1(\alpha R)J_0(\alpha r) - \frac{r}{R}J_0(\alpha R)J_1(\alpha r)}{J_1^2(\alpha R) - J_2(\alpha R)J_0(\alpha R)}v_0 e^{-\alpha z} + \left(\frac{2Q_0}{\pi R^2} - \frac{4v_0}{\alpha R}\right)\left(1 - \frac{r^2}{R^2}\right)$$ (6.78)

and

$$p(r,z) = p_0 + \frac{2v_0}{R}\left(1 - e^{-\alpha z}\right)\frac{J_0(\alpha R)}{J_0(\alpha R)J_2(\alpha R) - J_1^2(\alpha R)} - \frac{4z}{R^2}\left(\frac{2Q_0}{\pi R^2} - \frac{4v_0}{\alpha R}\right)$$ (6.79)

where we have assumed that the function $\phi(z)$, which is the value of v_r at $r = R$ is given by

$$\phi(z) = v_0 e^{-\alpha z}.$$ (6.80)

6.4 Flow Measurement by Indicator Dilution Method

6.4.1 *Introduction*

An indicator or a tracer is a substance which in some way labels a moving fluid so that the movement of the indicator is representative to that of the fluid which it labels. An indicator can be introduced into the living body and is not metabolised. The presence of an indicator does not alter any properties of the system and it can easily be recovered from the system. The distribution of an indicator in the system is considered to be a linear function of the input.

6.4.2 *Measurement of Flow*

Let us consider the case of a vascular system which has only a single arterial inlet and a single venous outlet. We shall assume that the indicator is a dye. Suppose an amount m of an indicator is injected into the blood stream on the venous side of the left ventricle. We assume that the ventricle behaves as a reservoir having the same rate of influx as that of outflux. This assumption obviously neglects the periodic nature of blood flow. It is further assumed that the blood in the veins containing the full amount of dye will enter the left ventricle and then proceed to the aorta. The concentration of dye in the aorta can be measured.

The method is based on the simple concept that a known amount of indicator is injected into the blood stream and the rate of flow is measured by the rate of dilution of the indicator. It is, therefore, based on the law of conservation of mass. A similar method, known as Allen's method, is used in fluid engineering for determining the volume of water in a conduit.

Although a variety of indicators, such as sodium, iodine, nitrous oxide, carbon monoxide, acetylene, etc. have been used, the most frequently used indicator is cardiogree dye or indocyanine green.

Let $c(t)$ be the concentration of dye in the left ventricle, V be the volume of blood flow in the left ventricle and k be the *Cardiac Output* (CO), the flux of blood through the left ventricle or the steady rate at which blood is ejected by the heart. The cardiac output is an important parameter in determining the condition of the heart with reference to its proper functioning.

We assume that all the dye moves with the blood from the left ventricle to the aorta, and no amount of dye can escape through any of the upstream arteries.

Let the time be measured from the instant when the dye is injected into a systemetic vein and t_0 be the time when the dye enters into the ventricle. The initial amount of dye that is injected and observed into the ventricle is

$$m = Vc(t_0). \tag{6.81}$$

We therefore have the basic equation for the concentration of dye from one-compartment model analysis (Mazumdar [68])

$$\frac{d}{dt}(Vc) = -kc, \quad t > t_0, \tag{6.82}$$

whose solution is given by

$$c(t) = \frac{m}{V}e^{-\frac{k}{V}(t-t_0)}, \quad t > t_0. \tag{6.83}$$

It is interesting to note that the actual experimental curve which is obtained by Zierler [128] shows the following facts: The time that the dye takes to reach the heart is approximately 5 secs. After this time delay, the concentration of dye initially increases linearly to a maximum value and then the curve behaves exponentially. The initial linear increase may be accounted for by reasons such as the time taken for the injection, the process of diffusion that takes place after the injection and the transport phenomenon for the dye to reach the left ventricle. However, if the experimental curve, say $\bar{c}(t)$, is represented mathematically after the initial linear phase of duration t_0 elapses, then we have

$$\bar{c}(t) = ae^{-bt} \tag{6.84}$$

where a and b are dimensional constants. Comparing Eqs. (6.83) and (6.84), we obtain the following relations

$$a = \frac{m}{V}e^{\frac{k}{V}t_0}, \quad b = \frac{k}{V}. \tag{6.85}$$

Calculating the area A under the experimental curve and comparing it with the theoretical value, we obtain

$$\begin{aligned} A &= \int_0^\infty \bar{c}(t)\,dt \\ &= \int_{t_0}^\infty c(t)\,dt \\ &= \frac{m}{V}\int_{t_0}^\infty e^{-\frac{k}{V}(t-t_0)}\,dt \\ &= \frac{m}{k}, \quad t > t_0. \end{aligned} \tag{6.86}$$

Thus, we obtain

$$k = \frac{m}{A} = \frac{m}{\int_0^\infty \bar{c}(t)\,dt},$$

$$V = \frac{m}{Ab},$$

$$t_0 = \frac{1}{b} \log\left(\frac{a}{Ab}\right). \tag{6.87}$$

The theoretical curve thus obtained with the value of t_0 obtained from the above is shown in Fig. 6.7.

Fig. 6.7: The theoretical curve of dye concentration.

However, the typical dye concentration curve recorded from a human subject looks like Fig. 6.8. Because of recirculation of dye, the concentration does not fall-off as in the theoretical curve. The dashed line on the descending curve represents the semilogarithmic extrapolation of the upper portion of the descending part, prior to the beginning of recirculation.

From Eq. (6.87), we conclude that the cardiac output k is given by

$$k = \frac{\text{mass of indicator injected}}{\text{area under the indicator-dilution curve}}. \tag{6.88}$$

This result is usually referred to as *Hamilton's formula*.

Fig. 6.8: A typical indicator dilution curve.

6.5 Peristaltic Flows

6.5.1 *Introduction*

The word peristalsis stems from the Greek word *peristaltikos*, which means clasping and compressing. Peristalsis is a muscle-controlled flow similar to the flow in the cardiovascular system. Peristalsis occurs in many organs. Peristaltic flow is the flow generated in the fluid contained in a distensible tube when a progressive wave travels along the wall of the tube. Although the elasticity of the wall does not directly enter into the flow equations, it affects the flow through the progressive wave travelling along its length. Peristaltic flow occurs widely in the functioning of ureter, transport of spermatozoa in the cervical canal, transport of bile, transport of cilia, vaso motion of small blood vessels, etc. Of these physiological flows, peristalsis in ureter, intestine and stomach are the most important, from the biofluid mechanics point of view. The mathematical problem of peristaltic flow is similar to that of the collapsible tube. In the case of functioning of the ureter, we model it mathematically assuming that the ureter receives fluid from the kidney at the upper end, and passes it down to the bladder against a pressure gradient. Normally, there is more than one wave along the entire length of the ureter, which is of the order of 30 cm. The amplitude of the wave is of the order of 5 mm and its speed is about 6 cm/sec. The frequency of contractions is about 1 to 8 per minute. Each contraction lasts about 1.5 to 9 sec, the diastolic (expansion) phase is about twice as long as the systolic (contraction) phase. Pressure during the contraction varies from 2 to 8 mm Hg at the pelvis, 2 to 10 mm Hg in the upper portion of the ureter, and 2 to 14 mm Hg in the lower portion (see Weinberg [117]).

6.5.2 *Peristaltic Motion in a Cylindrical Tube*

On the basis of peristaltic transport in the ureter described in Sec. 6.5.1, let us now consider an idealised two-dimensional peristaltic problem of studying the wave motion of the wall of a circular cylindrical tube at a moderate amplitude. A sinusoidal wave is assumed to travel down the walls of a two-dimensional channel of constant width and infinite length. We assume the cylindrical tube or the channel is filled with a homogeneous Newtonian viscous fluid, although the fluid involved may be non-Newtonian, and the flow may take place in two layers, i.e., a core layer and a peripheral layer. Taking the x-axis along the center line of the channel, and the y-axis normal to it, the equations governing two-dimensional motion of a viscous fluid, the continuity and Navier-Stokes equations, are given by (see Sec. 1.3)

$$\frac{\partial u}{\partial x} + \frac{\partial v}{\partial y} = 0, \tag{6.89}$$

$$\frac{\partial u}{\partial t} + u\frac{\partial u}{\partial x} + v\frac{\partial u}{\partial y} = -\frac{1}{\rho}\frac{\partial p}{\partial x} + \nu\nabla^2 u, \qquad (6.90)$$

$$\frac{\partial v}{\partial t} + u\frac{\partial v}{\partial x} + v\frac{\partial v}{\partial y} = -\frac{1}{\rho}\frac{\partial p}{\partial y} + \nu\nabla^2 v. \qquad (6.91)$$

If we use the stream function ψ as given in Eq. (1.36), i.e., $u = \psi_y$, $v = -\psi_x$, and eliminate the pressure terms in Eqs. (6.90) and (6.91) we obtain as in Eq. (1.38)

$$\nabla^2\psi_t + \psi_y\nabla^2\psi_x - \psi_x\nabla^2\psi_y = \nu\nabla^4\psi \qquad (6.92)$$

in which

$$\nabla^2 \equiv \frac{\partial^2}{\partial x^2} + \frac{\partial^2}{\partial y^2}. \qquad (6.93)$$

We assume the fluid is subjected to conditions imposed by the symmetric motion of the elastic walls. Let the vertical displacements of the upper and lower walls be η and $-\eta$, respectively. Hence, the equations of the walls are given by

$$y = \pm\eta(x,t) = \pm a\left[1 + \epsilon\cos\frac{2\pi}{\lambda}(x - ct)\right] \qquad (6.94)$$

where ϵ is the amplitude ratio, λ is the wavelength, c is the wave speed and a is the undeformed radius of the tube (see Fig. 6.9).

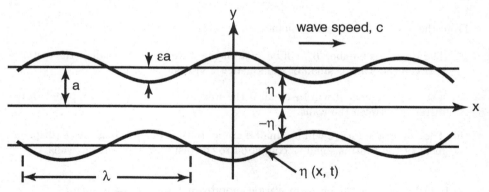

Fig. 6.9: Peristaltic motion of the flexible walls.

The main objectives in this study are to determine the longitudinal pressure gradient that can be generated by the travelling wave, and the flow resulting from peristalsis superimposed on pressure differences at the ends of the tube.

In order to solve Eq. (6.92) we assume the following boundary conditions

$$u = 0$$

and

$$v = \pm \frac{2\pi ac\epsilon}{\lambda} \sin \frac{2\pi}{\lambda}(x - ct) \quad \text{at} \quad y = \pm\eta(x,t). \tag{6.95}$$

Introducing the dimensionless quantities

$$X = \frac{x}{\lambda}, \quad Y = \frac{y}{a}, \quad T = \frac{ct}{\lambda}, \quad \Psi = \frac{\psi}{ac},$$

$$n = \frac{a}{\lambda}, \quad R_e = \frac{ac}{\nu}, \tag{6.96}$$

Eq. (6.92) finally reduces to

$$D_n^2 \Psi_T + \Psi_Y D_n^2 \Psi_X - \Psi_X D_n^2 \Psi_Y = \frac{1}{nR_e} D_n^2 \Psi \tag{6.97}$$

where the operator D_n is given by

$$D_n^2 \equiv n^2 \frac{\partial^2}{\partial X^2} + \frac{\partial^2}{\partial Y^2}. \tag{6.98}$$

The corresponding boundary conditions become

$$\Psi_Y = 0$$

and

$$\Psi_X = 2\pi\epsilon \sin 2\pi(X - T) \quad \text{at} \quad Y = \pm\eta(X, T). \tag{6.99}$$

From the above, dimensionless quantities, we notice the following:

(i) The Reynolds number R_e will be small if the wave speed is small or the distance between the walls is small or the kinematic viscosity is large.

(ii) The wave number n will be small if the wavelength is large as compared to the distance between the walls.

(iii) The amplitude ratio ϵ will be small if the amplitude of travelling waves (displacement of the wall) is small as compared to the distance between the walls.

In order to solve the problem, various approximations are made such as

(i) small Reynolds number so that nonlinear convective terms in the Navier-Stokes equation can be neglected,

(ii) long-wavelength compared with the undeformed radius of the tube,

(iii) small amplitude of the wall displacement compared with the tube radius.

6.5.3 *Long-Wavelength Analysis*

Let us make the assumptions of small Reynolds number and small wave number, i.e.,

$$R_e \ll 1, \quad n \ll 1. \tag{6.100}$$

Consider cylindrical coordinates such that the Z-axis is along the centre-line of the tube. The equation of the tube wall is given by

$$h(Z, t) = a \left(1 + \epsilon \sin \frac{2\pi}{\lambda}(Z - ct) \right) \tag{6.101}$$

where, as before, a is the undeformed radius of the tube (see Fig. 6.10). We assume that the pressure is independent of the radial coordinate, i.e.,

$$p = p(Z, t). \tag{6.102}$$

We introduce a moving coordinate system (r, z) travelling with the wave (see Kapur [50]) so that

$$r = R, \quad z = Z - ct. \tag{6.103}$$

We have the continuity equation and equation of motion as given before in Eqs. (1.44) and (1.46)

$$\frac{\partial}{\partial r}(ru) + \frac{\partial}{\partial z}(rw) = 0, \tag{6.104}$$

$$\frac{\partial p}{\partial z} = \mu \left(\frac{\partial^2 w}{\partial r^2} + \frac{1}{r} \frac{\partial w}{\partial r} \right) \tag{6.105}$$

where u and w are the radial and axial velocity components of the fluid in the moving coordinate system.

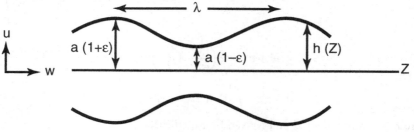

Fig. 6.10: The geometry of the tube.

Using the boundary condition

$$w = -c \quad \text{at} \quad r = h \tag{6.106}$$

we get from Eq. (6.105)

$$w = -c - \frac{1}{4\mu}\frac{\partial p}{\partial z}(h^2 - r^2). \tag{6.107}$$

The flow rate is given by

$$q = 2\pi \int_0^h rw\,dr$$

$$= -\pi h^2 c - \frac{\pi h^4}{8\mu}\frac{\partial p}{\partial z}. \tag{6.108}$$

So, we have

$$\frac{\partial p}{\partial z} = -\frac{8\mu q}{\pi h^4} - \frac{8\mu c}{h^2}. \tag{6.109}$$

We thus obtain from Eq. (6.107)

$$w = -c + 2\left(\frac{q}{\pi h^4} + \frac{c}{h^2}\right)\left(h^2 - r^2\right). \tag{6.110}$$

For radial velocity u, we integrate Eq. (6.104) and obtain

$$ru = -\int_0^r r\frac{\partial w}{\partial z}\,dr, \tag{6.111}$$

which with the help of Eq. (6.110) reduces to

$$u = -\frac{\partial h}{\partial z}\left(\frac{cr^3}{h^3} - \frac{2qr}{\pi h^3} + \frac{2qr^3}{\pi h^5}\right). \tag{6.112}$$

Reverting back to the stationary system, the velocity components U and W, and the flow rate Q are given by

$$U = u = -\frac{\partial h}{\partial Z}\left(\frac{cR^3}{h^3} - \frac{2qR}{\pi h^3} + \frac{2qR^3}{\pi h^5}\right)$$

$$= -\frac{2\pi a\epsilon}{\lambda}\cos\frac{2\pi}{\lambda}(Z - ct)\left(\frac{cR^3}{h^3} - \frac{2qR}{\pi h^3} + \frac{2qR^3}{\pi h^5}\right) \tag{6.113}$$

$$W = w + c = 2\left(\frac{q}{\pi h^4} + \frac{c}{h^2}\right)(h^2 - R^2) \tag{6.114}$$

$$Q = 2\pi \int_0^h WR\,dR = q + \pi ch^2 \tag{6.115}$$

where h is a function of Z and t given by Eq. (6.101).

To determine the pressure drop across one wavelength λ, we have from Eq. (6.109), after integrating,

$$(\nabla p)_\lambda = -\frac{8\mu q}{\pi a^4}\int_0^\lambda \frac{dz}{\left(1 + \epsilon\sin\frac{2\pi}{\lambda}z\right)^4} - \frac{8\mu c}{a^2}\int_0^\lambda \frac{dz}{\left(1 + \epsilon\sin\frac{2\pi}{\lambda}z\right)^2}$$

$$= -\frac{4\mu}{\pi a^4}\left[\frac{q(2+3\epsilon^2)}{(1-\epsilon^2)^{\frac{7}{2}}} + \frac{2\pi ca^2}{(1-\epsilon^2)^{\frac{3}{2}}}\right]. \tag{6.116}$$

Clearly, the pressure drop across one wavelength would be zero if

$$q = -\frac{2\pi ca^2(1-\epsilon^2)^2}{(2+3\epsilon^2)} \tag{6.117}$$

which gives the corresponding velocity components

$$U = -\frac{2\pi a\epsilon cR}{\lambda h^3}\left[R^2 + \frac{4ca^2(1-\epsilon^2)^2}{2+3\epsilon^2}\left(1 - \frac{R^2}{h^2}\right)\right]\cos\frac{2\pi}{\lambda}(Z - ct), \tag{6.118}$$

$$W = 2c\left(1 - \frac{2a^2(1-\epsilon^2)^2}{h^2(2+3\epsilon^2)}\right)\left(1 - \frac{R^2}{h^2}\right). \tag{6.119}$$

Clearly, when $\epsilon = 0$, the results are that of a cylindrical tube of uniform cross-section.

Chapter 7

FLUID MECHANICS OF HEART VALVES

7.1 General Introduction

The heart valves are mechanical devices that permit the flow of blood in one direction only. The two cuspid (atrio-ventricular) valves are located between the atria and ventricles, while the two semi-lunar valves are located at the entrance to the pulmonary artery and the great aorta. The cuspid valves prevent blood from flowing back into the atria from the ventricles and the semi-lunar valves prevent it from regurgitating into the ventricles from the aorta and pulmonary artery.

Any of the four valves may lose its ability to close tightly; such a condition is known as valvular incompetency or regurgitation. On the other hand, stenosis is an abnormality in which the fully-open valvular orifice becomes narrowed by scar tissue that forms as a result of valvular disease; thereby the blood flow through the valve results in increased pressure drop across the valve orifice.

Heart valve disease in advanced forms severely affects the performance of a natural heart valve by causing permanent alteration to the valve tissue leading to valve dysfunction. An improper functioning heart valve induces considerable disability and may ultimately lead to death if the problem is not treated. The need for surgical repair or total valvular replacement arises when a patient's quality of life suffers from the ensuing disability. Biological and prosthetic valve substitutes have been employed since the early 1960s when valve replacement was first introduced as a clinical procedure. Today, mechanical prostheses and tissue valves (bioprostheses) are in widespread clinical use in medical centres throughout the world.

Despite over 25 years of development and clinical experience prosthetic heart valves continue to suffer from numerous pathological problems. The ideal valve substitute has yet to be designed. Although many difficulties can be expected during and immediately following a valve implant, the most debilitating and life-threatening complications are those that begin to develop from the moment of implant and manifest themselves in latter years.

The presence of the prosthesis disturbs the flow of blood producing areas of high fluid stress, high wall stress and separated flow regions. Depending upon their extent, these flow conditions may eventually cause many serious pathological problems including sub-haemolytic injury and haemolysis, endothelial damage and erosion, thrombotic deposition and related thromboembolic events. There is a direct link between the fluid dynamics of heart valves and the pathological problems encountered. Hence, it is extremely important to examine the likelihood of these problems by investigating as fully as possible the fluid dynamics of prosthetic heart valves.

Measurements using *in vitro* experiments provide much valuable information. However, physical measurements are often limited in application due to probe size or are simply not suitable for many regions of the flow domain. This has been the experience of previous researchers in this field (Figiola and Mueller [23]; Herkes and Lloyd [39]). Experimental measurements are unable to provide a complete picture of the complex flow patterns.

Computational fluid dynamic techniques, that will be discussed in the next chapter, provide a powerful noninvasive tool for obtaining good approximations to velocity and stress distributions in a flow field. However, depending upon the nature of the fluid dynamics, the mathematical equations and the discrete modelling, the numerical solutions can range from very accurate to only rough approximations. Current numerical techniques provide acceptably good results for most physical flow problems. Only the more complex flow cases (e.g., turbulent pulsatile flow) cause difficulties. The advent of high speed digital computers and ready access to modern computing power has generated strong interest in the development of numerical models to investigate prosthetic valve fluid dynamics. A number of important studies in this field have for the most part concentrated on laminar flow and/or idealised valve geometries. However, in reality, a complex combination of both laminar and turbulent flow and its relaminarization is observed through prosthetic valves during the cardiac cycle. Furthermore, the tissue walls are generally curved.

In the next chapter, the application of computational methods to elucidate the complex fluid dynamics associated with steady flow through prosthetic heart valves, will be examined by including turbulence effects and accurate boundary shape.

7.2 A Brief Description of the Heart Valves

As we know, the human heart is a pulsatile synchronous pump which beats more than 40 million times each year. Its function is to provide the mechanism by which the oxygen requirements of the human body are satisfied by maintaining the circulation of blood.

It has been mentioned in Chap. 2, that the human heart has four chambers — two chambers pumping blood and two chambers receiving blood. The right and the left ventricles are the two separate pumping chambers of the heart, pumping blood to the lungs and the body through two large blood vessels. The vessel supplying the lungs is the pulmonary artery, and that supplying the body is the aorta. The two receiving chambers in the heart are the left and right atria. The one receiving oxygenated blood from the lungs is the left atrium, and that receiving deoxygenated blood from the veins of the body is the right atrium.

The human heart has four valves, two on each side. The valves on the right side are the tricuspid valve — an inlet valve, and the pulmonary valve — an outlet valve. The valves on the left are the mitral valve — an inlet valve, and the aortic valve — an outlet valve. In the natural heart, the valve leaflets open at the centre to allow unobstructed central flow. The action of the four heart valves within the heart acts to restrict blood flow to one direction and prevent substantial backflow.

The aortic and pulmonary valves each consist of three cusps attached to a fibrous tissue ring on the inner walls of the aorta and pulmonary artery, respectively. Both outlet valves have similar dimensions. The adult aortic or pulmonary valve is approximately 25 mm in diameter. Behind each cusp is a pouch-like cavity embedded into the artery wall. These regions are known as the sinuses of Valsalva. An important mechanism in aortic valve closure is the production of vortices in the region between each cusp and its sinus. Bellhouse and Talbot [8] demonstrated this behaviour by constructing a trileaflet valve mounted in a rigid chamber with a model aortic sinus.

The mitral and tricuspid valves are also similar to each other. However, the mitral valve has two major and two minor leaflets, whereas the tricuspid valve has three similarly sized leaflets. Unlike the other valves, the circumference of the mitral valve is not circular but "D-shaped", measuring approximately 35 mm across the major diameter. A cutaway view of the heart illustrating anatomical relationships is shown in Fig. 7.1.

After circulating through the body, deoxygenated blood is collected in the great veins. It then flows into the right atrium and through the tricuspid valve into the right ventricle. The right ventricle contracts, and together with the closure of the tricuspid valve, the deoxygenated blood is ejected through the pulmonary valve into the pulmonary artery. The blood flows to the lungs where the transfer of carbon dioxide and oxygen occurs. Four pulmonary veins then direct the reoxygenated blood into the left atrium where the blood flows through the mitral valve into the left ventricle. This filling phase of the cardiac cycle is known as diastole.

When the diastolic cycle is complete, the left ventricle contracts abruptly, the mitral valve closes, and blood is ejected through the aortic valve. This phase of the cardiac cycle is known as ventricular systole. Left and right ventricular systole occur simultaneously as do left and right diastole. The muscular contractions are controlled by a group of cells situated in the heart. The blood which is ejected through the aortic valve passes down the aorta into the arteries and then to the veins which reach most of the body. A schematic diagram of blood flow within the heart is shown in Fig. 7.2 (see also Fig. 2.3, p. 21).

Fig. 7.1: Drawing of the four cardiac valves as viewed from the base of the heart. Note the manner in which the leaflets overlap in the closed valves.

The left and right coronary arteries supply oxygenated blood to the muscle of the heart known as the *myocardium*. These arteries originate from two of the three sinuses of Valsalva located at the base of the aorta. The flow of blood through these arteries is crucial for the proper functioning of the heart muscle.

The volume of blood passing down the aorta during each systole equals the left ventricular stroke volume, minus the amount which flows into the coronary arteries, minus the amount of retrograde flow through the aortic and mitral valves. Cardiac output in man varies between about 5 litres min^{-1} at rest to about 25 litres min^{-1} during extreme exertion. Normally retrograde flow is negligible except in diseased valves where it may be considerable. The stroke volume may then become several times greater than the flow down the aorta. This is clearly undesirable as the cardiac output of a healthy individual is normally sufficient to fulfill the oxygen requirements of the body.

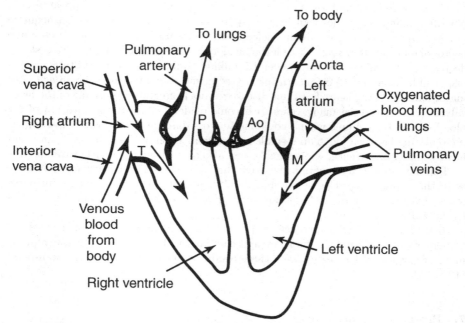

Fig. 7.2: Schematic diagram of the blood flow through the heart valves. The arrows show the direction of blood flow. The valves are indicated by symbols: T, tricuspid; P, pulmonary; Ao, aortic; and M, mitral.

A valve is called incompetent when it fails to close properly, causing a jet of blood to flow retrogradely through the valve when it should be shut. Both mitral and aortic incompetence are common valvular defects.

Another common valve defect is stenosis. This is a narrowing of the valve which restricts the flow of blood across the valve. Increased pressure drops are consequently induced across a stenosed valve. In mitral stenosis this pressure increase is transmitted back through the lungs to the pulmonary artery. The body may then try to compensate for these effects by reducing cardiac output making physical exertion difficult and uncomfortable. Irreversible damage to lung function and right ventricular failure have been known to occur as a result of prolonged severe mitral stenosis. Aortic stenosis also occurs causing a raised left ventricular systolic and a reduced aortic pressure. The effect of blood flowing through a stenosed aortic valve at high pressure is to cause turbulence in the aorta, aortic dilation and deposition of calcium on the aortic wall and on the valve leaflets. Tricuspid stenosis and incompetence often occurs in association with mitral valve stenosis and incompetence, respectively.

The human heart valves may become defective through acquired disease or because of congenital malformation. Congenital defects may initially appear minor, but after many years they can become quite serious. The most common acquired diseases which may cause valvular defects are rheumatic valvular heart disease and bacterial endocarditis. Even though the incidence of rheumatic heart disease is slowly decreasing, it remains present in over 1.5 million adults in the United States, and is responsible for about 15,000 deaths each year in subjects under 65 years of age. The major damage to the heart from this disease results from gradual scarring of the heart valves which may take many years to become evident. A bacterial invasion of the blood stream by a simple wound could cause gross damage to the heart valve leaflets if circulating bacteria lodge on a small valve lesion. Both of these diseases can result in valvular incompetence, while stenosis can also result from rheumatic valve disease.

Papillary muscle dysfunction can occur due to blockage of coronary artery branches to the papillary muscles of the mitral valve, resulting in sudden mitral incompetence. Even a myocardial infarct can sometimes injure the attachments of the valves to the wall.

It is clear that if incompetence or stenosis is present in heart valves, then serious consequences can ensue for the heart and circulation. The need for valvular replacement or surgical repair arises when these conditions threaten the life and well-being of the patient.

7.3 Prosthetic Heart Valves

The preservation of proper circulation is of paramount importance to patient well being, and this often necessitates the replacement of a dysfunctioning valve, or valves, by artificial substitutes. Today, nearly 75000 prosthetic heart valves are implanted annually throughout the world.

Let us examine many important aspects of prosthetic heart valves. In particular, the history of valve replacement, a general description of prosthetic valves, and the numerous pathological complications that may and very often do arise following a valve implant.

7.3.1 *History of Valve Replacement*

Prior to the development of prosthetic valves, the surgical correction of valvular defects carried a high mortality rate and any benefits obtained were usually only short-lived. With mitral stenosis, surgical treatment was directed towards conversion of the stenosis into incompetence. The first successful treatment was performed by

Souttar [101] in 1925, who used his fingers to dilate a stenotic mitral valve. Many years later the second successful treatment of this type was performed by Bailey [5] in 1949. By the early 1950s, closed valvulotomy, which consisted of blindly inserting the finger through the atrium or ventricle and relieving the stenotic valve in the beating heart, had become a safe and successful procedure. Dilatation of stenotic valves using fingers was almost entirely discontinued in 1959 with the introduction of the transventricular dilator. This mechanical dilator was appreciated for providing much less general disturbance to the heart.

A new era in cardiac surgery emerged with the development of the heart-lung machine for the extra-corporeal circulation of the blood. It had become possible to operate on the completely arrested open heart while the heart-lung machine served to supply oxygenated blood to the rest of the body. As a direct consequence of this development, a great number of new possibilities became available in valvular repair. The pathway for total valve replacement had been opened and the design of suitable valve replacements constituted a major cardiac research project during the 1960s. The primary consideration of the valve designers involved in the initial development was prevention of any serious regurgitation in the closed valve.

In 1953 the first prosthetic heart valve was implanted by Hufnagel without the use of a heart-lung machine (Hufnagel [40]). The actual insertion time required only between 3 and 6 minutes. The valve was of the caged-ball type and was placed into the descending aorta of patients for the treatment of severe aortic incompetence. Locating the valve in a more physiologic position was simply too difficult. Unfortunately, the valve was unable to prevent regurgitation from the aortic arch vessels and was also prone to a high incidence of thromboembolic events.

In 1960, Harken [37] performed the first subcoronary insertion of a caged-ball valve in a group of patients with aortic regurgitation. Although the caged-ball prosthesis was at first employed unsuccessfully in the aortic position it was successfully used for mitral valve replacement by Starr and Edwards [102] in the same year. Following a great number of modifications, the caged-ball prosthesis continues to remain in use today. However, its centrally obstructive design gives it only adequate haemodynamics and is responsible for thrombogenic events and mild haemolysis.

The caged-disk valve was introduced in 1965 by several investigators in an attempt to improve haemodynamics and provide a lower profile and lighter weight than the caged-ball valve (Hufnagel and Conrad [41]; Kay [52]). Unfortunately, the caged-disk design is the most flow obstructive of all prosthetic valves, particularly in the aortic position where it has been shown to increase pre-existing mitral insufficiency. Other problems include high thrombogenicity, haemolysis, intimal proliferation in the aortic root and disk cocking.

In an attempt to overcome these difficulties the tilting-disk prosthesis was introduced in 1967 by Wada. Lillehei-Kaster and Björk-Shiley tilting-disk valves were introduced soon after (Kaster-Lillehei [51]; Björk [10]). Initial problems with disk

wear caused by the hinge struts in the Björk-Shiley prosthesis were overcome by incorporating into the design the ability for the disk to turn freely once every 180 to
200 cycles. The potential for disk swelling was removed with the introduction of the
pyrolytic carbon disk in 1972. A recent series of the tilting-disk valve use a convex-
concave disk which allows greater forward flow by rising further out of its housing
when open. The semi-central nature of flow through tilting-disk valves and excellent
haemodynamics have made this valve a popular choice with cardiac surgeons [85].

St. Jude Medical, Inc. introduced a bileaflet pivoting-disk prosthesis for clinical
trial in 1977. The valve consists of two hinged semi-circular leaflets housed in an
orifice ring which have an opening angle of 85° and a closing angle of 30° to 35°. The
pivoting mechanism eliminates the need for the supporting struts designed into both
the Björk-Shiley and Lillehei-Kaster tilting-disk valves. When open the St. Jude
valve provides a superior high-flow orifice with nearly complete centralised flow. The
St. Jude prosthesis moved from clinical investigation status to clinical use in early
1983 with a complication rate as low or lower than that of any other mechanical valve
available.

Several cardiac surgeons have preferred to use biological substitutes for valve replacement. Homograft valves (removed from a cadaver), heterograft valves (removed
from a dead animal), or tissue-fabricated valves (constructed from nonvalvular human and animal tissue) have been used extensively for the replacement of diseased
aortic and mitral valves. Tissue-fabricated valves are designed to duplicate the flow
behaviour of natural valves. Although short-term results are good, the long-term
results for tissue bioprosthetic valves are not satisfactory due to calcification, leaflet
degeneration and eventual valve stenosis. The main advantage of employing tissue
valves over synthetic mechanical valves is their almost complete freedom from thrombus formation. To this date, no mechanical or biological valve replacement has been
developed which is completely satisfactory.

Nearly 50 different prosthetic heart valve designs have been developed since the
early 1960s. The vast majority of these valves have been discarded and, of those
remaining, several modifications have been made to improve haemodynamic performance, wear characteristics and overall valve function. At present there are five major
categories of prosthetic valves in use: (1) Caged-ball; (2) caged-disk; (3) tilting-disk;
(4) bileaflet pivoting-disk; and (5) tissue bioprosthesis.

A typical mechanical prosthesis consists of four basic components: (1) Occluder;
(2) cage, strut or hinge; (3) sewing ring; and (4) orifice ring. The occluder responds
passively to pressure and flow changes within the heart to allow forward flow and
to prevent substantial backflow. It is free to move under action of blood flow. The
cage, strut or hinge retains and guides the motion of the occluder. The prosthesis
is attached into the heart by sutures which pass through the sewing ring, into the
tissues and back through the sewing ring. Both the cage (or struts) and sewing ring
are attached to the orifice ring. Blood flows through the orifice when the valve is

open and is sealed by the occluder when the valve closes (see Fig. 7.3).

Fig. 7.3: Diagrammatic representation of major categories of mechanical prosthetic heart valves in use today. The valves are fully open and key components are identified.

There are numerous prosthesis-related complications that can arise following a valve implant [43]. Fatal and non-fatal problems which are unrelated to valve fluid dynamics include infection (prosthetic valve endocarditis), haemorrhage during anti-coagulation therapy, mechanical malfunction due to material degradation, separation of prosthesis from its attachment site (valve dehiscence) and interference with cardiac function.

Other non-fatal conditions related to valve fluid dynamics which may cause concern include haemolysis, sub-haemolytic injury, endothelial damage and tissue overgrowth. It is clear, therefore, that the fluid dynamics of prosthetic valves play a crucial role in the aetiology of several important prosthesis-related complications.

Let us now discuss a few prosthesis-related complications attributable to valve fluid dynamics.

7.3.2 *Thrombosis and Thromboembolism*

Thrombosis is defined as the formation of a blood coagulum within a vessel or the heart. There are several factors which favour thrombus formation. These include:

(1) Stasis, or reduction in velocity of blood flow, which is common in varicose veins and inactivity;

(2) derangement in composition of blood as may be experienced through disease, surgery or childbirth;

(3) recirculating, or separated flow caused by disruption to normal flow by the presence of a flow constriction as in a heart valve prosthesis; and

(4) alterations in the endothelium which may be due to inflammation, vascular diseases or elevated wall shear stresses.

Such flow conditions prevail at various locations in the vicinity of heart valve prostheses. During systole, a predilection site for thrombotic deposition occurs near the flow reattachment point of the sinus recirculation zone and along the downstream face of the sewing ring. In the large circulation zone, the downstream side of an occluder-type valve (e.g., caged-disk or caged-ball) is also a region of potential thrombus formation. The presence of a cage or strut further enhances the possibility of thrombogenesis. Varying amounts of thrombus formation have been observed on recovered Starr-Edwards aortic caged-ball valves with a predominance at the base of the cage struts and the apex of the cage. A schematic example of thrombus formation and tissue overgrowth is shown in Fig. 7.4.

Fig. 7.4: Diagrammatic example of thrombus formation and tissue overgrowth on a Starr-Edwards caged-ball valve.

The complications that can arise from thrombosis of a prosthetic valve are:

(1) *Thrombotic stenosis*, when the prosthesis is partially blocked with a thrombus thus inhibiting the valve's performance; and

(2) *Thromboembolic events*, when part of the thrombus, known as an embolus, breaks off the main thrombus body and is carried by the blood to lodge in the small arteries of a vital organ such as the heart, kidney or brain causing temporary or permanent damage.

The incidence of thromboembolism from prosthetic valves represents a major threat to patients. Consequently, intense anticoagulation therapy is mandatory with any mechanical valve implant.

7.3.3 *Haemolysis*

The basic characteristics of shear flow apply not only to pure fluids but also to complex suspensions like blood. Locally accelerated blood flow occurs readily in the vicinity of prosthetic heart valves inducing abnormally high fluid forces, or shear stress. A blood cell which is subjected to a region of shearing action exerted by the surrounding fluid will experience a distribution of shear stress over its entire membrane. Consequently, the cell membrane will be stretched and may suffer irreversible changes detrimental to its essential function.

Sub-haemolytic injury, or damage to blood cells, is common in prosthetic valve recipients but its effects can usually be compensated by the body's ability to substantially increase its rate of haemoglobin production. If the rate of destruction exceeds the rate of production, clinically significant anaemia will be observed. The common cause of haemolytic anaemia is the development of a leak between the sewing ring and its attachment site. This leak is physically similar to jet-like flow and induces extremely high shearing forces on the blood. The prosthesis itself continues to damage blood cells by its mere presence in the flow field, with significant anaemia usually absent.

Estimates for shear induced in-bulk haemolysis thresholds have varied widely, from 150 to over 4000 Nm^{-2}. Several reasons for these apparent discrepancies have been proposed including the extent to which containing surfaces are responsible for the observed levels of haemolysis. Leverett *et al.* [59] proposed that shear stress and exposure time are the two primary determinants for in-bulk haemolysis. Cells exposed for very short durations to very high stresses do not lyse, but exposure for long durations at much lower stress levels can cause lysis. However, below a certain threshold of fluid stress normal red blood cells will not deform and rupture even if exposed for an indefinite period. Two independent studies, one performed with turbulent shear flow (Sutera and Mehrjardi [106]) and the other with laminar shear flow (Leverett *et al.* [59]) have shown this low stress threshold is between 150 and 250 Nm^{-2}. It can also be inferred that the presence of the fluctuating characteristics of turbulent flow are not harmful provided the time-averaged shear stress does not exceed the threshold.

7.3.4 *Endothelial Damage*

The endothelium is a delicate membrane which lines the heart, blood vessels and body cavities. In large arteries such as the aorta, rapidly flowing blood exerts a force, or viscous drag, upon the endothelial lining of the artery which attempts to pull the lining along with the flow. This force, or shear stress, is directly proportional to the velocity gradient near the vessel wall. Normally the levels of wall shear stress are so small that no damage or erosion of endothelial cells occurs. However, during systole, the peripheral flow induced by caged-disk and caged-ball valves, and the semi-central flow through tilting-disk prostheses causes a significant increase in velocity gradients at the proximal portions of the aorta. As has already been mentioned, damage to the endothelium also acts to increase the potential for thrombus formation by increasing the "stickiness" of the vessel wall. It is therefore clear that a highly desirable feature of any valve prosthesis should be low levels of wall shear stress.

7.3.5 *Tissue Overgrowth*

Tissue overgrowth is often seen on valve prostheses recovered during autopsy. Yoganathan [126] observed tissue overgrowth on the aortic side of the sewing ring and on the struts of cloth-covered Starr-Edwards caged-ball valves (see Fig. 7.4). The cloth covering (e.g., polypropylene mesh) on cage struts and valve orifice has reduced the incidence of thromboembolism by allowing infiltration of fibrous tissue which eventually serves as a nonthrombogenic surface. However, excessive tissue growth into the cloth covering will eventually result in valve stenosis and an increase in the pressure drop across the valve, thus diminishing hydraulic performance. The development of less thrombogenic materials (e.g., pyrolytic carbon) for use in valve construction has led to some progress in the control of tissue overgrowth. Nevertheless, tissue overgrowth is not limited to cloth covered valves where it has been primarily noticed, but appears to depend to some extent on the fluid mechanics of the valve, although this process is not fully understood. High blood flow and the associated effects of fluid scouring may play a role in limiting or preventing tissue overgrowth.

Chapter 8

COMPUTATIONAL BIOFLUID MECHANICS

8.1 Introduction

The development of computational methods to solve nonlinear partial differential equations has been progressing for over 60 years. The advent of high speed digital computers has seen many advances in this field by enabling development, comparison and refinement of the numerical techniques. A historical outline of computational fluid dynamics is available in Roache [86]. Publications of Journals such as *Journal of Computational Physics, Computers and Fluids, Journal of Fluid Mechanics* and *Numerical Methods in Engineering* contain papers describing the application of numerical methods to numerous physical problems including physiological flow processes.

One particular method, the *finite difference method*, remains the predominant technique for obtaining numerical solutions to fluid flow problems. The method can be summarised by the following stages (Noye [79]):

(1) A grid is developed to cover the solution region. At the grid points (i.e., intersection of grid lines) the approximate solution to the problem will be found.

(2) All dependent variables are initialised. The initial values are either known exactly, or they may at best be only a rough guess, especially if they require determination through application of the solution procedure.

(3) The finite difference equations corresponding to the continuous partial differential equations are constructed. They are used to calculate approximate values of the dependent variables at all grid points in the solution domain.

(4) The computational cycle commences. New values are calculated at adjoining space positions by incrementing some value Δx, and at subsequent time levels by Δt. The process is continued until a predetermined time is reached, or as in the steady state, the dependent variables at each grid point remain unchanged at successive time levels.

Unfortunately, it is generally impossible to obtain theoretical estimates of the accuracy, convergence and stability of a numerical solution to a problem possessing large nonlinear terms in the governing equations. This is especially true for viscous fluid flow problems with large Reynolds numbers. In order to verify the accuracy of finite difference techniques applied to such problems, it is necessary to compare numerical solutions with experimental results and/or employ computational experimentation such as observing solution behaviour when subject to grid refinement. The mathematical theory for convergence and stability of numerical schemes have only been developed for linear systems. Consequently, results from linear theory are used to guide nonlinear problems but many *ad hoc* approaches remain a practical necessity. For example, with fluid flow problems, it is often necessary to anticipate the nature of the solution in advance so that a larger number of grid points can be assigned to regions of interest, particularly those with high velocity gradients. Typically, these regions occur at constrictions and near wall boundaries. Also, specification of iterative relaxation parameters normally requires some testing beforehand to determine optimal values. It may be found that switching to slightly different values after a certain number of global iterations serves to enhance convergence and control stability. Alternating the direction of iterative sweeps also seems to provide additional solution control. Monitoring solution progress is mandatory and its implementation depends upon the nature of the problem, the availability of experimental results for comparison and the degree of interactiveness of the computing environment. In this regard, computer graphics are very useful.

The vast majority of problems associated with computational fluid dynamics are solved using finite difference methods and a regular Cartesian coordinate system. Consequently, flow domain geometry is often simplified by making boundary walls coincide with the rectangular computational mesh. In a sense, the physical domain is fitted to the computing mesh by adjusting the location of grid lines until the flow domain is satisfactorily defined. Clearly, this approach is not suited for application to more general problems, particularly those where curved physical boundaries strongly influence the flow characteristics. Domains with greater geometrical complexity are often solved with the more formidable finite element method and demand great programming effort when compared to the finite difference approach. A technique that embodies the computational efficiency of finite differences and facilitates application to general body problems is the numerical generation of *boundary-fitted coordinates*.

In a boundary-fitted coordinate system, the physical domain (z, r) is numerically transformed, or mapped, onto a calculation domain in coordinated (ξ, η) — a regular rectangular coordinate system (Fig. 8.1). All numerical calculations involving the governing equations are conveniently performed in the transformed (ξ, η) plane rather than the physical (z, r) plane. Complicated geometries exhibiting much complex structure may best be handled using a composite grid.

The boundary-fitted system approach also offers a substantial advantage over the usual treatment of curved boundaries with finite differences. For example, a rectangular mesh may have been overlayed on any general shape and either a "saw tooth" approximation or interpolation could be employed for the curved regions. It is reasonable to expect that the boundary-fitted coordinate system would provide a better boundary representation in such areas. Other advantages of the boundary-fitted approach include:

(1) Simplification of boundary condition specification;

(2) accurate implementation of boundary conditions;

(3) efficient use of grid points for the calculation area; and

(4) flexibility in the type of curved-wall problems that can be addressed.

Once the physical domain is specified, a suitable boundary-fitted computing mesh can be generated. Unfortunately, this is seldom an easy task. The reason for this arises partly from the difficulty in providing a sufficiently good initial estimate for the grid and in the choice of an appropriate numerical procedure. Thompson and Warsi [110] provide a comprehensive review of numerical methods for generating curvilinear coordinate systems having coordinate lines coincident with boundary segments.

8.2 Mathematical Modelling

Modelling pulsatile flow through the circulatory system in precise mathematical terms remains an insurmountable problem. The system is affected by many physical and chemical factors. In the case of prosthetic heart valves, the situation is further complicated by the presence of an unnatural device in the flow path. In spite of the difficulty created by the multiplicity of interrelated elements, extensive insight into fluid flows can still be obtained by applying numerical solution techniques to simpler physical systems.

The literature contains many examples of mathematical models for a variety of systems (physical, biological, chemical, etc. . . .) with analytic or numerical methods of

Fig. 8.1: Example of a non-rectangular geometry in the physical plane with schematic grid lines and its corresponding representation in the transformed plane.

solution available. Consequently, this chapter is not concerned in detailing the many complex equations which make up fluid mechanics but instead restricts the discussion to a relatively qualitative level. This chapter describes some basic fluid laws relating to flow through blood vessels and prosthetic heart valves. Mathematical models of turbulence are discussed. Computational methods are covered for solving nonlinear mathematical flow equations and generating boundary-fitted coordinate systems.

8.3 Laminar Versus Turbulent Flow Models

As we have seen earlier, there are two basic modes of viscous fluid motion — laminar and turbulent flow. If we consider axisymmetric flows in a pipe, then in laminar flow particles follow straight lines parallel to the pipe axis, but in turbulent flow the particles follow a random path through the pipe, and only the average motion is along the pipe axis. Laminar and turbulent flows are essentially states of stability and instability, respectively, and as such, no complete definition of turbulence seems possible.

As explained earlier, experimental observations indicate that above a critical Reynolds number, transition will occur from laminar to turbulent flow. Thus, laminar flow can become unstable. Below a certain Reynolds number, small disturbances introduced to the flow are always damped out and the flow remains stable. The random fluctuating motion characteristic of turbulent flow arises from small disturbances which are not damped out, but instead increase with time and eventually interact with each other in a chaotic manner. As a result, differences between laminar and turbulent flow become evident. Firstly, the velocity profiles appear very different. The fully developed laminar profile is parabolic and the turbulent profile is much flatter near the pipe centre with larger gradients near the vessel walls. Secondly, the "roughness" of turbulent flow acts to increase the effective viscosity of the fluid. In the laminar case, the viscosity is entirely due to molecular effects. However, for turbulent flow the apparent viscosity comes from two sources:

(i) molecular viscosity; and

(ii) convection of turbulent "eddies" which usually form the larger part of the effective viscosity. This additional term is often referred to as the *eddy viscosity*.

In the circulatory system, the flow of blood is normally laminar. However, abrupt flow variations due to irregularities and stenoses, as are often encountered with prosthetic heart valves, will produce turbulence. Experimental research indicates that turbulent flow can occur during the period of ejection in the ascending aorta of subjects with elevated cardiac output, and it occurs consistently in the ascending aorta of subjects with abnormal aortic valves or with prosthetic aortic valves (Stein and

Sabbah [103]). It is apparent from the discussion above that turbulence will cause a substantial increase in the level of shear stress since both the effective viscosity and the shear strain rate will become larger, particularly near a vessel wall. Given valve geometry, fluid properties and details of the inflow conditions one would prefer to evaluate in totality the velocity and shear stress fields for the entire flow domain.

The connection between the observed complexity of fluid flow and the fundamental physical laws governing such behaviour is usually obscure. Indeed, with what we know, both the basic laws and experimental observation are required to make the connection. Unfortunately, a mathematical description of fluid flow based on physical laws has not yet developed sufficiently so as to cater for every possible flow situation. Even in cases where the mathematics is known, the formulation is often posed in highly nonlinear terms, thus excluding analytic solution. Similarly, the implementation of experimental measurements is time-consuming and expensive, and the accuracy of observations, as limited as they may be, is often questioned.

Nevertheless, the need to address real-life problems is pressing, and mathematical models of fluid flow, both laminar and turbulent, have been developed and applied successfully to many problems. These models have governing partial differential equations which are based on the well-known conservation principles of mass, momentum and energy. The mathematical formulation is usually nonlinear and analytic application to realistic problems is often very difficult or impossible unless simplifications are made. By exploiting the capabilities of modern high-speed computers and numerical techniques, solutions to otherwise unsolvable problems are being found.

The following two sections provide a discussion of turbulence models and the numerical methods that have been applied to fluid flow problems.

8.4 Turbulence Models

Turbulence modelling of complex engineering and physical flows has been a topic of active research for many years. Several mathematical models have been developed, and these are usually classified as either Reynolds stress models or eddy viscosity models. Reynolds stress models are based on the evolutionary equations for the turbulent stresses. They were developed to overcome deficiencies in eddy viscosity models. Although successful in computing free shear flows, they have encountered difficulties in treating wall-bounded separated flows. In addition, their complex structure results in increased computer time and decreased stability. It is questionable whether the benefits outweigh the costs. As a result, Reynolds stress models have been applied to relatively few turbulent flow problems (see Rodi [88]).

Eddy viscosity models, however, are much more popular. These models are based on the *Kolmogorov-Prandtl* proposal that the eddy viscosity is proportional to the product of a turbulent velocity scale and a turbulent length scale. That is

$$\nu_t \propto qL \tag{8.1}$$

where ν_t is the eddy viscosity, q and L are velocity and length scales of the turbulence, respectively. Depending upon the manner in which the eddy viscosity is specified, these models can be further grouped into either algebraic or differential models. Algebraic models ν_t are determined by replacing q and L with simple algebraic expressions involving the mean velocity field. Mixing-length models, as they are also known, are excellent for investigating single-zone flows but become deficient when switching between one type of zone to another within a single flow. This is particularly noticeable when investigating separated flows.

Differential models specify the eddy viscosity in the solution of one or two differential equations. One-equation models prescribe the turbulent length scale but use a partial differential equation to determine the turbulent kinetic energy. The kinetic energy serves to characterise the velocity scale. This approach results in only a marginal improvement over the mixing-length models.

Two-equation models employ the Boussinesq eddy-viscosity concept

$$\overline{u_i' u_j'} = -\nu_t \left(\frac{\partial U_i}{\partial x_j} + \frac{\partial U_j}{\partial x_i} \right) + \frac{2}{3} k \delta_{ij} \tag{8.2}$$

where δ_{ij} is the Kronecker delta, and U_i and U_j are the mean components of velocity in the i and j directions, respectively. Differential equations are used to involve both the velocity and length scales. One of the most studied two-equation turbulence models is the $k - \epsilon$ *model*. This model remains the most widely used approach for the solution of practical turbulent flow problems. In this model, the eddy viscosity is obtained by solving partial differential transport equations for the turbulent kinetic energy k and the energy dissipation rate ϵ. These transport equations are relatively easy to incorporate into existing Navier-Stokes computer codes. The complete equations are presented in the subsequent sections.

The $k - \epsilon$ model has been successfully applied to a great number of shear flow problems. However, for more complicated flows the model tends to lose its accuracy. The standard $k - \epsilon$ model is based on the assumption that the eddy viscosity is identical for all Reynolds stresses. In two-dimensional flows without swirl this condition has proved adequate for the calculation of thin shear layers. However, in certain three-dimensional flow situations (e.g., square duct flows) and in planar separated flows the isotropic eddy viscosity assumption proves insufficient by underpredicting velocity fluctuations, and thus giving rise to inaccurate predictions for the Reynolds stresses. The normal Reynolds stresses play an important role in the calculation of the mean velocity field in separated turbulent flows. For example, it is well known that the $k - \epsilon$ model underpredicts the recirculation length down-stream of a backward-facing step. Another consequence is the inability to predict secondary flows in non-circular ducts. A refinement which uses an algebraic stress model can be introduced to replace both the Kolmogorov-Prandtl expression (8.1) and the Boussinesq eddy-viscosity relation (8.2) above (Rodi [88]).

A limitation on the $k - \epsilon$ turbulence model is the restriction to high Reynolds number flows (i.e., $R_e > 20000$). Consequently, it is not directly applicable to the viscous sublayer near boundary walls and to low Reynolds number flows. However, the common approach of using wall functions allows efficient treatment of the near wall regions. In the case of low Reynolds number flows the standard $k - \epsilon$ model and wall-function method are insufficient and alternative approaches have been de-veloped. The most popular of these is the low-Reynolds number turbulence model of Jones and Launder [47]. This model seeks to provide turbulence equations which are valid throughout the laminar, semi-laminar and fully turbulent regions. The costs associated with this improvement are the requirement for many additional compu-tations in the near wall region, increased equation complexity and related numerical instability.

Nevertheless, the $k - \epsilon$ turbulence model and wall-function method is the simplest approach being widely applied to practical turbulent flow problems. Once imple-mented, it is the stepping stone by which the model refinements mentioned above may eventually be introduced.

8.5 Computational Methods for the Study of Flow Through Prosthetic Heart Valves

As explained in Chap. 7, the *in vitro* measurement of flow through heart valve prostheses produces useful but limited information, and in some instances, such as the potential for valve thrombosis, very little can be deduced from a finite number of measuring points. As a result, extensive clinical trials are required to establish whether prosthetic valves have acceptably low levels of blood trauma and thrombo-sis. Computational fluid dynamic techniques, however, provide a powerful tool for determining valve haemodynamic performance as they can yield detailed quantitative data for velocity and stress distributions in the flow field. Thus, conclusions regarding the potential for a valve to induce blood trauma or give rise to thrombotic deposition can be drawn via computational methods. Additional benefits include reliability, ef-ficiency, immediacy and ready adaption to various geometric and flow configurations. Nevertheless, computational techniques and experimental measurements will continue to complement each other as they currently do in many aspects of modelling in other physical sciences.

Several authors have employed numerical methods to determine blood flow patterns in the vicinity of prosthetic heart valves. Au and Greenfield [1] applied numerical techniques and principles of fluid dynamics to study stress distribution in blood caused by the axial motion of an occluder in a caged-disk prosthesis. They assumed planar geometry, laminar Newtonian flow and a comparatively low Reynolds number of 50 (based upon distance between the occluder and cage strut, and the occluder velocity). Time-dependent finite difference equations were formulated and an interactive graphics environment provided continuous monitoring of numerical procedure for stability and convergence rates. Changes in maximum stress values as the occluder moved from the fully closed to an almost fully open position were reported. However, the emphasis was on the application of interactive computer graphics to study this and other medical phenomena.

Au and Greenfield [2, 3] again utilised finite difference methods to examine steady laminar flow through a planar model of the Björk-Shiley tilting-disk prosthesis. The accent was again on the totality of the graphics displays which included contour maps and surface plots of computer-formed streamline, velocity, stress and pressure fields. Potential sites for haemodynamically induced thrombogenesis were observed. Importantly, the difficulty associated with computing on irregular boundary segments was noted and it was suggested that the finite element approach could overcome this problem.

Underwood and Mueller [112] obtained numerical solutions for steady axisymmetric laminar flow through a disk-type valve in a constant diameter chamber for Reynolds numbers ranging from 20 to 1300. The finite difference method was used to compute flow variables and the solution procedure includes calculation of a suitable time-step at each interval. Experimentation with both first-order and second-order accurate upwind differences for the advection terms revealed variation in the numerical results — the second method provides better solutions when compared with physical measurements. Regions of separated flow attached to the disk occluder, downstream of the sewing ring and along the aortic wall were identified. Zones of high fluid stress were also noted. The possibility of thrombosis and haemolysis was discussed.

In an extension of the above study, Underwood and Mueller [113] repeated the calculations, but instead of assuming a constant diameter chamber a more realistic aortic-shaped vessel was considered. A non-uniform grid system was used for both the axial and radial directions enabling close approximation of the aortic wall shape. Differences in solutions between the two chambers were reported. Reynolds numbers examined were between 50 and 600. Regions of separated flow were again present, particularly a recirculating zone on the chamber wall downstream of the sinus of Valsalva. This region was initially observed at a Reynolds number of about 200 and grew with increasing flow rates.

Engelman *et al.* [22] performed comparative studies on the steady laminar flow characteristics of Starr-Edwards caged-ball and Björk-Shiley tilting-disk valves in the mitral and aortic positions. The finite element method was used to model planar non-Newtonian fluid flow through the prostheses. The location of major eddies were identified in velocity vector plots for both valve types. Simulations using a Newtonian model were also carried out with markedly different results to the non-Newtonian model being reported for each configuration case.

Peskin [80] developed a computational method whereby the mathematical problem is formulated in such a way that the heart muscle and valves appear as special regions of the flow where extra forces apply. These forces give the valves and tissue walls their material and physiological properties. He assumed laminar Newtonian time-dependent flow. The flow equations were coupled to a nonlinear system of equations describing boundary motion and forces. Unfortunately, the method was restricted to very low Reynolds numbers ($\approx 1/25$ of the appropriate value) due to impractical computational demands. The study illustrated the use of computational methods to optimise prosthetic heart valve design and encouraged the use of the method to examine valve performance when subjected to changes in the physiological state of the heart (e.g., exercise).

Yang and Wang [123, 124] applied the $k - \epsilon$ turbulence model to investigate the problem of steady turbulent flow through a disk-type prosthetic heart valve in a constant diameter chamber. They predicted streamline, velocity, stress and pressure fields by solving a set of simultaneous algebraic finite difference equations using a line-iteration technique. Unlike laminar flow studies which can have up to four separated flow regions in a constant diameter chamber and three separated regions in an aortic-shaped chamber, the turbulent study uncovered only two such regions. Specifically, attached to the downstream side of the sewing ring and the disk occluder — there was no recirculation zone present along the chamber wall. High levels of fluid stress and wall shear stress were also evident.

Idelsohn *et al.* [44] carried out a comparative study on steady laminar flow through three common prosthetic heart valves. Using the finite element method, velocity, stress and pressure fields were calculated for caged-ball, caged-disk and tilting-disk prostheses in an aortic-shaped chamber. Regions of accelerated and stagnant flow were presented with the aid of computer graphics. Also, the location and magnitude of maximum shear stresses were found. Although these positions were generally in good agreement with experimental measurements, the levels of stress were much lower than those observed in physical simulators. The discrepancy between the numerical and experimental values indicated the important contribution of turbulent shear stresses.

Stevenson and Yoganathan [104, 105] conducted a numerical investigation for steady turbulent flow through axisymmetric models of trileaflet tissue aortic heart valves. Body conforming grid systems were generated and the turbulent flow equations were solved using finite differences and successive over-relaxation. They computed the flow patterns for five valve geometries ranging from very stenotic to a natural heart valve model. Good agreement was reported between experimental measurements and simulation data for velocity, shear stress and pressure drops. Their numerical model provided much useful information and may become a valuable tool for the development of future tissue prostheses.

8.6 Laminar Flow Model Through a Prosthesis

Let us consider a model for steady laminar flow through the idealised planar geometry of a disk-type valve. Although very simple, this model clearly demonstrates the potential effectiveness and usefulness of more sophisticated numerical and mathematical models. The method of finite differences is used to solve the model. Numerical solutions are provided for several Reynolds numbers.

8.6.1 *Problem Formulation*

The mathematical modelling of blood flow through the complex geometry of a prosthetic heart valve is a difficult task. In such a problem the complex geometries of the valve must be simplified so that they can be studied numerically. The simplifications must still contain the essential features of the valve. An approximate solution to the realistic problem may then be deduced from the study performed on the simplified model with the necessary assumptions taken into account. Thus, an insight into the real-life problem can be obtained.

The present analysis is performed on a disk-type prosthetic heart valve which can be used in either the mitral or aortic position, see Mazumdar and Thalassoudis [67]. In this study the valve is assumed to be in the aortic position. The sinuses of Valsalva and the structure of the valve cage will undoubtedly influence the flow field near an aortic valve. However, the occluder and the sewing ring will remain the overwhelming factors influencing the flow characteristics of the valve. Thus, it is decided not to include the sinuses and the valve cage in the present simple model. For the purpose of mathematical modelling, the laminar incompressible two-dimensional steady flow of a homogeneous Newtonian fluid with constant viscosity is assumed. The flow is considered during the greater part of systole when the valve is fully open. This is the worst case in terms of the extent of flow separation and the magnitude of fluid stresses.

A simple geometry is employed for which experimental data is available. The essential features present in this model are the sewing ring, the disk occluder, the aortic wall and a Poiseuille inflow velocity profile. The flow is assumed to be symmetric about the centreline, and hence only one half of the actual geometry is used for the analysis. A definition sketch for the disk valve model is given in Fig. 8.2. The dimensions of the valve are given in Fig. 8.3.

Based upon the foregoing assumptions, the governing equations for two-dimensional steady incompressible flow of a Newtonian fluid with no body forces are the

Fig. 8.2: Geometry and inflow velocity profile for the disk valve.

Fig. 8.3: Dimensions (cm) of the fully open disk valve.

continuity equation and the Navier-Stokes equations (i.e., Eqs. (1.33)–(1.35))

$$\frac{\partial u}{\partial x} + \frac{\partial v}{\partial y} = 0, \tag{8.3}$$

$$u\frac{\partial u}{\partial x} + v\frac{\partial u}{\partial y} + \frac{\partial p}{\partial x} = \frac{1}{R_e}\left(\frac{\partial^2 u}{\partial x^2} + \frac{\partial^2 u}{\partial y^2}\right), \tag{8.4}$$

$$u\frac{\partial v}{\partial x} + v\frac{\partial v}{\partial y} + \frac{\partial p}{\partial y} = \frac{1}{R_e}\left(\frac{\partial^2 v}{\partial x^2} + \frac{\partial^2 v}{\partial y^2}\right) \tag{8.5}$$

where u and v are the horizontal and vertical velocity components, respectively, and p is the pressure. The Reynolds number R_e is defined as

$$R_e = \frac{D\overline{U}}{\nu} \tag{8.6}$$

where $\overline{U}$ is the average inflow velocity, D is the channel width and ν is the kinematic viscosity of the fluid.

By introducing the stream function $\psi(x,y)$ and the vorticity function $\omega(x,y)$ as defined in Chap. 1

$$u = \frac{\partial \psi}{\partial y} \quad \text{and} \quad v = -\frac{\partial \psi}{\partial x} \tag{8.7}$$

and

$$\omega = \frac{1}{2}\left(\frac{\partial v}{\partial x} - \frac{\partial u}{\partial y}\right) \tag{8.8}$$

we finally obtain the *vorticity-transport equation* and *Poisson's equation*,

$$\frac{\partial \psi}{\partial y}\frac{\partial \omega}{\partial x} - \frac{\partial \psi}{\partial x}\frac{\partial \omega}{\partial y} = \frac{1}{R_e}\left(\frac{\partial^2 \omega}{\partial x^2} + \frac{\partial^2 \omega}{\partial y^2}\right), \tag{8.9}$$

$$\frac{\partial^2 \psi}{\partial x^2} + \frac{\partial^2 \psi}{\partial y^2} = -2\omega. \tag{8.10}$$

The nonlinearity of the coupled partial differential equations and the disk valve geometry make it impossible to obtain a solution through an analytic approach. Instead, the use of finite difference techniques provides a way by which an approximate numerical solution for the problem may be obtained. Once a numerical solution becomes available the velocity components can be calculated using a discretisation of Eq. (8.7). The shear stress defined by

$$\sigma_{xy} = -\frac{1}{R_e}\left(\frac{\partial u}{\partial y} + \frac{\partial v}{\partial x}\right) \tag{8.11}$$

can then be obtained through a further discretisation, as can the horizontal normal stress

$$\sigma_{xx} = -\frac{2}{R_e}\left(\frac{\partial u}{\partial x}\right). \tag{8.12}$$

The boundary conditions imposed are:

(1) *Upstream boundary*

   A fully developed Poiseuille velocity profile for channel flow is assumed at the inflow boundary. Flow in large arteries is found to assume such profiles. To ensure a horizontal inflow, the vertical velocity component v is set to zero. The inflow stream function and vorticity are subsequently computed from Eqs. (8.7) and (8.8).

(2) *Centreline*

   The stream function is given an arbitrary value of zero along the centreline. The values of the vorticity and the vertical velocity component are also equated to zero along this axis of symmetry.

(3) *Solid walls*

   The stream function is given a constant (non-zero) value along the entire solid boundary which includes the sewing ring and the aortic wall. Along the occluder boundary the stream function is equated to zero. The no-slip boundary conditions ensure that both velocity components are zero in these locations. The vorticity on solid walls is calculated from Eqs. (8.7) and (8.8) by taking a Taylor series expansion of the stream function out from the wall.

(4) *Corner points*

   The value of the stream function at corner points deserves no special treatment and is taken as a constant value. The vorticity at a convex corner point is calculated as an average of two values obtained by considering the x and y directions separately. The vorticity in concave corner points is equated to zero.

(5) *Downstream boundary*

   The outflow boundary conditions employed by Friedman [25] for the stream function and vorticity are also used here. These conditions are

$$\frac{\partial \psi}{\partial x} = 0 \quad \text{and} \quad \frac{\partial \omega}{\partial y} - R_e \frac{\partial \psi}{\partial y}\left(\omega + \frac{1}{2}\frac{\partial^2 \psi}{\partial y^2}\right) = 0 \tag{8.13}$$

which satisfy the requirements for horizontal outflow and constant pressure far downstream. The distance of the outflow boundary conditions from the disk valve is considered to be sufficiently far so as not to severely influence the flow upstream.

8.6.2 *Finite Difference Formulation*

Before the mathematical model can be formulated into a set of finite difference equations, a computational mesh must be constructed. For simplicity a regularly spaced mesh is chosen and boundary locations are made to coincide with the nearest grid point. The dimensions of the grid spacing in x and y are denoted by Δx and Δy, respectively, and are both equal to 0.04 cm. The computational mesh is a 164×33 grid (5412 grid points). During the initial code development and testing several other grid densities were investigated. Comparison between the numerical solutions indicated some ambiguous results occurring at the lower grid densities. These grids were discarded. A section of the final grid system is shown in Fig. 8.4.

The governing Eqs. (8.9) and (8.10) are now written in finite difference form. There are numerous ways in which this discretisation can be achieved. For this simple model the method employed by Friedman [26] is used as the basis for the finite difference formulation. Defining H and K for convenience as

$$H = \psi_{i+1,j} - \psi_{i-1,j}, \qquad K = \psi_{i,j+1} - \psi_{i,j-1} \tag{8.14}$$

the advection terms in the vorticity transport equation (8.9) are approximated using first order accurate upwind differences if $H \geq 0$ and $K \geq 0$. That is,

$$\left.\frac{\partial \omega}{\partial x}\right|_{i,j} \approx \frac{\omega_{i,j} - \omega_{i-1,j}}{\Delta x} \tag{8.15}$$

and

$$\left.\frac{\partial \omega}{\partial y}\right|_{i,j} \approx \frac{\omega_{i,j+1} - \omega_{i,j}}{\Delta y} \tag{8.16}$$

if $H \geq 0$ and $K \geq 0$.

The remaining first derivatives of ψ and second derivatives representing the diffusion terms are approximated by the well-known central space approximations. Substitution into Eq. (8.9) and rearranging yields the following finite difference equation

$$\omega_{i,j} = \frac{\omega_{i+1,j} + \omega_{i-1,j} + \omega_{i,j+1} + \omega_{i,j-1} + \frac{R_e}{2}(H\omega_{i,j+1} + K\omega_{i-1,j})}{4 + \frac{R_e}{2}(H + K)} \tag{8.17}$$

if $H \geq 0$ and $K \geq 0$. Similar finite difference equations are formulated for the other three possibilities, namely

$$\omega_{i,j} = \frac{\omega_{i+1,j} + \omega_{i-1,j} + \omega_{i,j+1} + \omega_{i,j-1} + \frac{R_e}{2}(H\omega_{i,j+1} - K\omega_{i+1,j})}{4 + \frac{R_e}{2}(H + K)} \tag{8.18}$$

if $H \geq 0$ and $K < 0$,

$$\omega_{i,j} = \frac{\omega_{i+1,j} + \omega_{i-1,j} + \omega_{i,j+1} + \omega_{i,j-1} + \frac{R_e}{2}(K\omega_{i-1,j} - H\omega_{i,j-1})}{4 + \frac{R_e}{2}(K - H)} \tag{8.19}$$

Fig. 8.4: Finite difference grid (164 × 33) and grid nomenclature for the disk valve.

if $H < 0$ and $K \geq 0$, and

$$\omega_{i,j} = \frac{\omega_{i+1,j} + \omega_{i-1,j} + \omega_{i,j+1} + \omega_{i,j-1} - \frac{R_e}{2}(K\omega_{i+1,j} + H\omega_{i,j-1})}{4 - \frac{R_e}{2}(K + H)} \tag{8.20}$$

if $H < 0$ and $K < 0$. The presence of R_e in both the numerator and denominator in the equations above overcomes the cell Reynolds number restriction often associated with solving such fluid flow problems.

Boundary conditions are also discretised. A difficulty of any computational viscous fluid flow problem which employs the stream function-vorticity method is that the values of vorticity on the no-slip solid boundaries are not known in advance but are required to solve the discretised equations. To overcome this difficulty Eq. (8.8) is used to provide an initial estimate for the wall vorticity once the initial values for the stream function are specified. The wall vorticity at the grid point (i, j), denoted by $\omega_{i,j}$, is given by the second order accurate expression

$$\omega_{i,j} \approx \frac{3(\psi_{i,j} - \psi_{i+1,j})}{(\Delta x)^2} - \frac{\omega_{i+1,j}}{2} \tag{8.21}$$

when the solid wall is facing in the positive x-direction. Similar expressions are used for the other possible configurations.

8.6.3 *Numerical Solution*

The numerical procedure employed by Friedman provides a way by which a numerical solution to a steady laminar flow at high Reynolds numbers can be obtained using a relatively small amount of computer time. The procedure is used in a step by step manner to calculate both the stream function and the vorticity as follows:

(i)   Set initial values for $\psi^{(1)}$ and $\omega^{(1)}$.

(ii) Use Successive Over Relaxation (SOR) with a relaxation factor of 1.8 only once to calculate the stream function at all the grid points immediately interior to the solid boundaries using the new vorticity values.

(iii) Calculate the stream function at all remaining interior grid points using the new values determined in Step (ii) and the most recent vorticity values.

(iv) Return to Step (ii) and repeat the process until a convergent solution is obtained. Denote the convergent result by $\overline{\psi}^{(k+1)}$.

(v)   Set
$$\psi^{(k+1)} = 0.96\,\overline{\psi}^{(k+1)} + 0.04\,\psi^{(k)}. \tag{8.22}$$

(vi) Use the Gauss-Seidel iteration method only once to calculate the boundary vorticity values using the new stream function values.

(vii) Calculate the vorticity at all interior grid points using the new values determined in Step (vi) and the latest stream function values.

(viii) Return to Step (vi) and repeat the process until a convergent solution is obtained. Denote the convergent result by $\overline{\omega}^{(k+1)}$.

(ix) Set

$$\omega^{(k+1)} = 0.3\,\overline{\omega}^{(k+1)} + 0.7\,\omega^{(k)}. \tag{8.23}$$

(x)   Return to Step (ii) and repeat the process until both

$$\left|\psi^{(k+1)} - \psi^{(k)}\right| < \epsilon_1 \tag{8.24}$$

and

$$\left|\omega^{(k+1)} - \omega^{(k)}\right| < \epsilon_2 \tag{8.25}$$

are satisfied.

The stream function was iterated using the SOR method until the difference between its value at consecutive iterate levels was less than 0.0005. The vorticity function was iterated with the Gauss-Seidel method until the difference between its value at consecutive iterate levels was less than 0.2% of the maximum vorticity value. A requirement was that the convergence criteria be satisfied at every grid point. The global convergence conditions were set by assigning values to ϵ_1 and ϵ_2 of 0.0005 and 0.2% of the maximum vorticity value, respectively. Final global convergence was declared when Eqs. (8.24) and (8.25) were both satisfied 30 times.

Convergent numerical solutions are obtained for Reynolds numbers $R_e = 50$, 500, 1000 and 6000. Stream function, vorticity, horizontal velocity, vertical velocity and shear stress solutions are computed at every grid point. The numerical solutions presented are of course subject to artificial viscosity errors introduced by the numerical procedure. This should be kept in mind when comparisons are made with experimental studies, particularly in regions of high velocity diagonal to grid lines.

Stream function contour plots are presented in Fig. 8.5 for $R_e = 50$ to 6000. The extent of recirculating regions downstream of the disk occluder as well as upstream and downstream of the sewing ring are clearly seen. The location of the separation point on the occluder remained at the downstream corner for all Reynolds numbers tested. The large recirculating zone behind the occluder contained horizontal reverse flow velocities up to 1.57 times ($i = 80, j = 1$) the average inflow velocity for $R_e = 6000$. Interesting features in Fig. 8.5 are the relative dimensions of the closed streamlines in the separated flow region immediately downstream of the occluder. As the Reynolds number increases from 50 to 6000, the streamlines, $\psi = -0.05, -0.10$ and -0.15 become markedly smaller, until the one at the innermost ($\psi = -0.05$) eventually vanishes for $R_e > 1000$. Physically, the flow velocity inside the separated zone decreases relative to the surrounding flow as the Reynolds number increases.

A recirculating region of large proportions also appeared along the aortic wall downstream of the occluder (Fig. 8.5). Although it was not present at $R_e = 50$, this region initially appeared at about $R_e = 150$ and grew rapidly with higher Reynolds numbers. The separation point for this region was observed to move upstream as the Reynolds number increased. The largest value of wall shear stress occurring within this zone was $-4.1 \ \text{Nm}^{-2}$ ($i = 80, j = 33$) for $R_e = 6000$.

Fig. 8.5: Computed streamlines for $R_e = 50$, 1000 and 6000.

In Fig. 8.6 plots of the computed horizontal velocity profiles at various cross sections of the flow field are shown. The changing structure of the horizontal velocity profiles at different locations downstream of the valve can be seen as the Reynolds number increases from 50 to 6000. They show the presence of the slow moving reversed flow region lying along the aortic wall. The potential for thrombus formation in this stagnated flow area is clearly enhanced, particularly if the endothelium has been damaged. It is interesting to note that whereas the $R_e = 50$ horizontal velocity profile in Fig. 8.6b almost re-estalished the parabolic profile it had at inflow, the same cannot be said for the $R_e = 6000$ case. The great flow disturbance that occurs at this high Reynolds number persists far downstream.

The distribution of wall shear stress along the aortic wall is given in Fig. 8.7 for $R_e = 50$ and 6000. Large variations in the wall shear stress occur as a result of lateral flow (induced by the valve geometry) impinging upon the vessel boundary.

(a) $i = 112$

(b) $i = 152$

Fig. 8.6: Computed horizontal velocity profiles for $R_e = 50$ and $R_e = 6000$ along various grid lines.

The "sharp" corners of the sewing ring structure also influence the wall shear stress levels. Points of flow separation are indicated in Fig. 8.7 by zero shear stress. The maximum levels of wall shear stress computed upstream of the sewing ring remained below 0.7 Nm^{-2} for $R_e = 6000$. The location of peak wall shear stress remained fixed ($i = 61, j = 33$) for all Reynolds numbers. The shear stress at this point was 61.8 Nm^{-2} for $R_e = 6000$. This is sufficient to damage endothelial cells, adhered red blood cells and platelets. However, the point of peak wall stress coincides with the location of the sinuses of Valsalva which are not considered in this simple model. In reality, the presence of these hollowed-out areas in the aortic root region would reduce the amount of stress, possibly averting extensive damage. The levels of wall shear stress diminished rapidly downstream from this point for all Reynolds numbers. The level of shear stress computed along the sewing ring reached a maximum value of 44.4 Nm^{-2} ($i = 47, j = 21$) for $R_e = 6000$. The maximum level of shear stress in the entire flow domain occurred just below the upstream corner of the occluder reaching 82.8 Nm^{-2} ($i = 59, j = 23$) for $R_e = 6000$.

The laminar shear stresses computed in this model appear likely to cause damage to adhered red blood cells, endothelial cells and platelets in several regions of the flow domain. Damage to red blood cells in flow is unlikely to occur based on the present results. However, when flow becomes turbulent, the turbulent stresses which are always much larger than viscous stresses could haemolyse blood cells in the vicinity of solid surfaces.

The potential for thrombus formation exists near the regions of separated flow observed at all flow rates. Thrombosis may occur in the recirculating region immediately behind the occluder, attaching itself to the restrictive cage, possibly inhibiting the mechanical performance of the disk valve. Thromboplastin may also be released from blood cells in the separated region lying alongside the aortic wall to enhance clotting. These conditions are clearly undesirable.

Although these results are obtained quite easily from a simple laminar model, they do provide a good illustration of the detailed descriptions of shear stress and velocity that are possible in the near vicinity of prosthetic heart valves. Both the flow patterns and the stress distributions indicate areas of concern and potential danger. However, it should be emphasised that turbulence will occur at the higher Reynolds numbers. The subsequent models incorporate the effects of turbulence.

Fig. 8.7: Computed wall shear stress distributions (dynes cm^{-2}) for $R_e = 50$ and $R_e = 6000$.

8.7 Turbulent Flow Model Through a Ball Prosthesis

8.7.1 *Introduction*

As an extension of the previous model, let us consider another model which examines steady turbulent axisymmetric flow through a Starr-Edwards caged-ball valve in an aortic-shaped chamber. A boundary-fitted coordinate system is generated for the flow domain in the vicinity of the valve. The coordinate lines follow the left ventricular wall, the sewing ring, and idealised sinus, the aortic wall and the occluder shape. The standard $k - \epsilon$ turbulence model is formulated using a stream function-vorticity approach. The governing partial differential equations are recast into their curvilinear equivalents before being discretised into finite difference equations which are then solved iteratively (Thalassoudis *et al.* [108]). The main aims of this turbulent model are to obtain quantitative estimates for velocity, wall shear stress and Reynolds stress distributions in the flow field surrounding the prosthetic valve.

8.7.2 *Model Formulation*

The geometry representing a fully open Starr-Edwards 1260-11A caged-ball prosthetic heart valve in an aortic-shaped chamber is shown in Fig. 8.8. The geometry includes a left ventricular chamber, a sewing ring structure, and idealised axisymmetric sinus, the aortic wall and a ball occluder. The dimensions of the valve are given in Fig. 8.9. For comparison purposes the dimensions are very close to those employed by Figliola and Mueller [23] in their experimental measurements. Assumptions made are that blood behaves like a Newtonian fluid in large arteries, the cage containing the occluder has a negligible effect upon the flow, the domain boundaries are smooth and rigid and the flow is axisymmetric. The fluid is given a density of 1.06 gm cm^{-3} and a molecular, or laminar, viscosity of 3.5×10^{-3} N sec m^{-2} closely matching that of human blood.

The analysis of turbulent flow is more difficult than the case of the laminar flow model discussed before, and it has become customary to abandon the study of the detailed flow structure and treat the flow as consisting of mean and fluctuating components. Using this approach the governing mass and momentum equations are decomposed through application of time-averaging. This procedure results in additional terms $\overline{\rho u_i' u_j'}$ often referred to as turbulent, or Reynolds stresses where ρ is the fluid density, u_i' and u_j' are the fluctuating components of velocity in the i and j directions, respectively. Closure of the model equations is achieved through the Boussinesq eddy-viscosity concept which assumes that the turbulent stresses are directly proportional to the mean velocity gradients. In general form this is expressed by

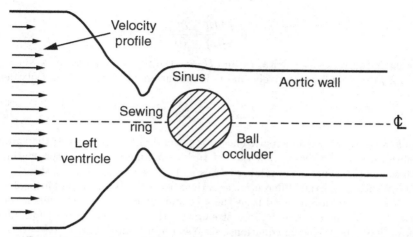

Fig. 8.8: Geometry and inflow velocity profile for the ball valve model.

Fig. 8.9: Dimensions (cm) of the fully open ball valve.

$$\overline{u_i' u_j'} = -\nu_t \left(\frac{\partial U_i}{\partial x_j} + \frac{\partial U_j}{\partial x_i} \right) + \frac{2}{3} k \delta_{ij} \tag{8.26}$$

where δ_{ij} is the Kronecker delta, U_i and U_j are the mean components of velocity in the i and j directions, respectively, and ν_t is the turbulent kinematic viscosity defined by

$$\nu_t = C_\mu \frac{k^2}{\epsilon} \tag{8.27}$$

where k is the turbulent kinematic energy representing the intensity of the turbulent fluctuations, ϵ is the dissipation rate of turbulent kinetic energy representing the length scale of the energy containing eddies and C_μ is an empirical constant.

The resulting partial differential equations together with the additional terms required for closure are referred to as the $k - \epsilon$ turbulence model. For this model the governing equations are written in a stream function-vorticity ($\psi - \omega$) formulation. Thus, we have the following equations (see Yang and Wang [124])

$$\omega = -\frac{1}{r} \left(\psi_{zz} + \psi_{rr} - \frac{1}{r} \psi_r \right), \tag{8.28}$$

$$\frac{1}{r} \left(\psi_r \omega_z - \psi_z \omega_r + \frac{\omega}{r} \psi_z \right) = (\nu_{eff}\omega)_{zz} + (\nu_{eff}\omega)_{rr} + \frac{1}{r}(\nu_{eff}\omega)_r - \frac{1}{r^2}\nu_{eff}\omega \tag{8.29}$$

where the effective kinematic viscosity ν_{eff} is defined as

$$\nu_{eff} = \nu + \nu_t \tag{8.30}$$

where ν is the molecular kinematic viscosity. The additional equations for the turbulent kinetic energy and its dissipation rate for a steady axisymmetric flow necessary to complete the model are

$$\frac{1}{r}(\psi_r k_z - \psi_z k_r) = \left[\left(\frac{\nu_{eff}}{\sigma_k} k_z \right)_z + \frac{1}{r} \left(r \frac{\nu_{eff}}{\sigma_k} k_r \right)_r \right] + G - C_D \epsilon, \tag{8.31}$$

$$\frac{1}{r}(\psi_r \epsilon_z - \psi_z \epsilon_r) = \left[\left(\frac{\nu_{eff}}{\sigma_\epsilon} \epsilon_z \right)_z + \frac{1}{r} \left(r \frac{\nu_{eff}}{\sigma_\epsilon} \epsilon_r \right)_r \right] + \frac{\epsilon}{k}(C_1 G - C_2 \epsilon). \tag{8.32}$$

The generation of turbulent kinetic energy G from the mean flow is given by

$$G = \nu_t \left\{ 2 \left[U_z^2 + V_r^2 + \left(\frac{V}{r} \right)^2 \right] + (U_r + V_z)^2 \right\} \tag{8.33}$$

where U and V, the mean axial and radial velocities, respectively, are defined by

$$U = \frac{1}{r}\psi_r \quad \text{and} \quad V = -\frac{1}{r}\psi_z. \tag{8.34}$$

$C_1, C_2, C_D, C_\mu, \sigma_k$ and σ_ϵ are empirical constants at high Reynolds numbers (Launder and Spalding [58]) whose values are given by

$$C_1 = 1.44, \quad C_2 = 1.92, \quad C_D = 1.0, \quad C_\mu = 0.09, \quad \sigma_k = 1.0, \quad \sigma_\epsilon = 1.3.$$

The governing equations are subject to several boundary conditions. At inflow we assume a fully developed turbulent velocity profile for pipe flow given by (Schlichting [97])

$$U = U_0 \left(1 - \frac{r}{R}\right)^{\frac{1}{7}}, \quad V = 0 \tag{8.35}$$

where U_0 is the mean axial centreline velocity at inflow, and R the inflow chamber radius. Analytic expressions for the inflow stream function and vorticity can be obtained from Eq. (8.35).

The turbulent kinetic energy and its dissipation rate at inflow are approximated by

$$k = 0.03\, U^2, \tag{8.36}$$

$$\epsilon = \frac{C_D k^{\frac{3}{2}}}{\lambda} \tag{8.37}$$

where λ is the macro eddy length scale. The numerical solutions proved insensitive to differing inflow conditions for ψ, ω, k and ϵ with the large blocking effect of the valve dominating the flow conditions in the vicinity of the prosthesis.

At outflow the boundary condition $\Phi_z = 0$ ($\Phi = \psi, \omega, k, \epsilon$) is imposed, and along the centreline the conditions are $\Phi_r = 0$ ($\Phi = k, \epsilon$), $\psi = 0$ and $\omega = 0$. The position of the outflow boundary at a distance of $10\,R$ past the ball occluder ensured the outflow conditions had little effect upon the flow upstream. Along solid boundaries the usual no-slip conditions are applied (i.e., $U = 0$ and $V = 0$) and the values of k and ϵ are also set to zero. The usual 'law of the wall' conditions are specified on the first grid point away from the wall. That is

$$\frac{U_p}{U^*} = \frac{1}{\kappa} \ln\left(E y^+\right), \tag{8.38}$$

$$y^+ = \frac{y U^*}{\nu}, \tag{8.39}$$

$$k = \frac{(U^*)^2}{\sqrt{C_\mu}}, \tag{8.40}$$

$$\epsilon = \frac{(U^*)^3}{\kappa y}, \tag{8.41}$$

$$\omega = \frac{U^*}{\kappa y} \tag{8.42}$$

where U_p is the velocity parallel to the wall, U^* is the friction velocity, κ is von Karman's constant ($\kappa = 0.4187$), E is van Driest's constant ($E = 9.793$), y^+ is the local Reynolds number and y is the grid point distance from the wall.

The velocity parallel to the wall at the first grid point away from the wall is computed along with the grid point distance from the boundary. An initial estimate for the friction velocity is made by assuming $U^* \approx |U_p|$. Based upon this estimate, y^+ is then evaluated. If $y^+ \leq 0.15$, the values of k and ϵ are equated to zero and the value of ω calculated from Eq. (8.28) is used. This approach is effective in controlling the stability of the numerical scheme during the early iterations, especially in the vicinity of separation and reattachment points where y^+ would occasionally assume very low values. For the vast majority of grid points adjacent to the boundary U^* is determined through a simple iterative process. If it is found that $y^+ > 11.63$, i.e., turbulence effects dominate, then k, ϵ and ω are computed from the equations above. Otherwise, if $y^+ \leq 11.63$ the flow is assumed to be purely laminar and the friction velocity is given by $U^* = y^+$ before k, ϵ and ω are computed. The sign of the vorticity computed above is assigned depending upon the sign of the corresponding U_p.

All the governing partial differential equations are recast into curvilinear coordinates by substituting derivatives in the (z, r) physical plane by their equivalent expressions in the (ξ, η) transformed plane.

Equations (8.28) and (8.29) are transformed into (ξ, η) coordinates to obtain

$$\omega = -\frac{1}{r}\left[\frac{1}{J^2}(\alpha\psi_{\xi\xi} - 2\beta\psi_{\xi\eta} + \gamma\psi_{\eta\eta} + \sigma\psi_\eta + \tau\psi_\xi) - \frac{1}{Jr}(z_\xi\psi_\eta - z_\eta\psi_\xi)\right] \tag{8.43}$$

and

$$\frac{1}{Jr}\left[\psi_\eta\omega_\xi - \psi_\xi\omega_\eta + \frac{\omega}{r}(r_\eta\psi_\xi - r_\xi\psi_\eta)\right]$$

$$= \frac{1}{J^2}\{\alpha(\nu_{eff}\omega)_{\xi\xi} - 2\beta(\nu_{eff}\omega)_{\xi\eta}$$

$$+ \gamma(\nu_{eff}\omega)_{\eta\eta} + \sigma(\nu_{eff}\omega)_\eta + \tau(\nu_{eff}\omega)_\xi\}$$

$$+ \frac{1}{Jr}\{z_\xi(\nu_{eff}\omega)_\eta - z_\eta(\nu_{eff}\omega)_\xi\} - \frac{\nu_{eff}\omega}{r^2} \tag{8.44}$$

where

$$J = z_\xi r_\eta - z_\eta r_\xi, \quad \alpha = z_\eta^2 + r_\eta^2, \quad \beta = z_\xi z_\eta + r_\xi r_\eta, \quad \gamma = z_\xi^2 + r_\xi^2, \tag{8.45}$$

$$\sigma = \frac{r_\xi(\alpha z_{\xi\xi} - 2\beta z_{\xi\eta} + \gamma z_{\eta\eta}) - z_\xi(\alpha r_{\xi\xi} - 2\beta r_{\xi\eta} + \gamma r_{\eta\eta})}{J} \tag{8.46}$$

and

$$\tau = \frac{z_\eta(\alpha r_{\xi\xi} - 2\beta r_{\xi\eta} + \gamma r_{\eta\eta}) - r_\eta(\alpha z_{\xi\xi} - 2\beta z_{\xi\eta} + \gamma z_{\eta\eta})}{J}. \tag{8.47}$$

For an orthogonal coordinate system $\beta = 0$.

We further have,

$$\phi_z \equiv \left(\frac{\partial\phi}{\partial z}\right)_{r,t} = (r_\eta\phi_\xi - r_\xi\phi_\eta)/J, \tag{8.48}$$

$$\phi_r \equiv \left(\frac{\partial\phi}{\partial r}\right)_{z,t} = (z_\xi\phi_\eta - z_\eta\phi_\xi)/J, \tag{8.49}$$

$$\phi_{zz} \equiv \left(\frac{\partial^2\phi}{\partial z^2}\right)_{r,t} = (r_\eta^2\phi_{\xi\xi} - 2r_\xi r_\eta\phi_{\xi\eta} + r_\xi^2\phi_{\eta\eta})/J^2$$
$$+ \left[(r_\eta^2 r_{\xi\xi} - 2r_\xi r_\eta r_{\xi\eta} + r_\xi^2 r_{\eta\eta})(z_\eta\phi_\xi - z_\xi\phi_\eta)\right.$$
$$\left.+ (r_\eta^2 z_{\xi\xi} - 2r_\xi r_\eta z_{\xi\eta} + r_\xi^2 z_{\eta\eta})(r_\xi\phi_\eta - r_\eta\phi_\xi)\right]/J^3, \tag{8.50}$$

$$\phi_{rr} \equiv \left(\frac{\partial^2\phi}{\partial r^2}\right)_{z,t} = (z_\eta^2\phi_{\xi\xi} - 2z_\xi z_\eta\phi_{\xi\eta} + z_\xi^2\phi_{\eta\eta})/J^2$$
$$+ \left[(z_\eta^2 r_{\xi\xi} - 2z_\xi z_\eta r_{\xi\eta} + z_\xi^2 r_{\eta\eta})(z_\eta\phi_\xi - z_\xi\phi_\eta)\right.$$
$$\left.+ (z_\eta^2 z_{\xi\xi} - 2z_\xi z_\eta z_{\xi\eta} + z_\xi^2 z_{\eta\eta})(r_\xi\phi_\eta - r_\eta\phi_\xi)\right]/J^3, \tag{8.51}$$

$$\phi_{zr} = \left[(z_\xi r_\eta + z_\eta r_\xi)\phi_{\xi\eta} - z_\xi r_\xi\phi_{\eta\eta} - z_\eta r_\eta\phi_{\xi\xi}\right]/J^2$$
$$+ \left[z_\eta r_\eta z_{\xi\xi} - (z_\xi r_\eta + z_\eta r_\xi)z_{\xi\eta} + z_\xi r_\xi z_{\eta\eta}\right](r_\eta\phi_\xi - r_\xi\phi_\eta)/J^3$$
$$+ \left[z_\eta r_\eta r_{\xi\xi} - (z_\xi r_\eta + z_\eta r_\xi)r_{\xi\eta} + z_\xi r_\xi r_{\eta\eta}\right](z_\xi\phi_\eta - z_\eta\phi_\xi)/J^3 \tag{8.52}$$

and

$$\nabla^2\phi = (\alpha\phi_{\xi\xi} - 2\beta\phi_{\xi\eta} + \gamma\phi_{\eta\eta} + \sigma\phi_\eta + \tau\phi_\xi)/J^2, \tag{8.53}$$

$$\vec{\nabla}\phi = \left[(r_\eta\phi_\xi - r_\xi\phi_\eta)\vec{i} + (z_\xi\phi_\eta - z_\eta\phi_\xi)\vec{j}\right]/J, \tag{8.54}$$

$$\frac{1}{Jr}(\psi_\eta k_\xi - \psi_\xi k_\eta) = \left(\frac{\nu_{eff}}{\sigma_k}\right)\left[\frac{1}{J^2}(\alpha k_{\xi\xi} - 2\beta k_{\xi\eta} + \gamma k_{\eta\eta}\right.$$

$$+\sigma k_\eta + \tau k_\xi) + \frac{1}{Jr}(z_\xi k_\eta - z_\eta k_\xi)\Bigg]$$

$$+\frac{1}{J^2}(r_\eta k_\xi - r_\xi k_\eta)\left\{r_\eta\left(\frac{\nu_{eff}}{\sigma_k}\right)_\xi - r_\xi\left(\frac{\nu_{eff}}{\sigma_k}\right)_\eta\right\}$$

$$+\frac{1}{J^2}(z_\xi k_\eta - z_\eta k_\xi)\left\{z_\xi\left(\frac{\nu_{eff}}{\sigma_k}\right)_\eta - z_\eta\left(\frac{\nu_{eff}}{\sigma_k}\right)_\xi\right\}$$

$$+G - C_D\epsilon \tag{8.55}$$

and

$$\frac{1}{Jr}(\psi_\eta \epsilon_\xi - \psi_\xi \epsilon_\eta) = \left(\frac{\nu_{eff}}{\sigma_\epsilon}\right)\left[\frac{1}{J^2}(\alpha \epsilon_{\xi\xi} - 2\beta \epsilon_{\xi\eta} + \gamma \epsilon_{\eta\eta}\right.$$

$$+\sigma\epsilon_\eta + \tau\epsilon_\xi) + \frac{1}{Jr}(z_\xi \epsilon_\eta - z_\eta \epsilon_\xi)\Bigg]$$

$$+\frac{1}{J^2}(r_\eta \epsilon_\xi - r_\xi \epsilon_\eta)\left\{r_\eta\left(\frac{\nu_{eff}}{\sigma_\epsilon}\right)_\xi - r_\xi\left(\frac{\nu_{eff}}{\sigma_\epsilon}\right)_\eta\right\}$$

$$+\frac{1}{J^2}(z_\xi \epsilon_\eta - z_\eta \epsilon_\xi)\left\{z_\xi\left(\frac{\nu_{eff}}{\sigma_\epsilon}\right)_\eta - z_\eta\left(\frac{\nu_{eff}}{\sigma_\epsilon}\right)_\xi\right\}$$

$$+\frac{\epsilon}{k}(C_1 G - C_2\epsilon). \tag{8.56}$$

The generation of turbulent kinetic energy G term given by Eq. (8.33) is also transformed to (ξ, η) coordinates.

Finally the Reynolds number for this model is defined by

$$R_e = \frac{DU_0}{\nu} \tag{8.57}$$

where D is the inflow chamber diameter and U_0 is the mean axial centreline velocity at inflow.

8.7.3 *Coordinate System Generation*

In the previous section, the geometric configuration of the caged-disk model is simple and the boundary walls are coincident with the Cartesian coordinate system. In this section, details are presented on the approach used to generate an orthogonal boundary-fitted coordinate system for a fully open Starr-Edwards caged-ball prosthesis.

The method requires that the shape of physical domain be known in advance and not be determined as part of the solution. Initially, the upper surface of the boundary shape shown in Fig. 8.8 was constructed from a combination of trigonometric functions for the left ventricular chamber, sinus of Valsalva and aortic wall, and cubic splines for the sewing ring. A hemisphere was used for the ball occluder (as only half the valve geometry is necessary due to the symmetry assumption) but this was altered slightly by "smoothing" the fore and aft sections of the hemisphere to reduce singularity effects which became evident during the computation process. The boundary functions for the upper part of the physical domain are:

(1) Inflow chamber, $0.0 \leq z < 4.0$:

$$r(z) = 3.2. \tag{8.58}$$

(2) Left ventricular section, $4.0 \leq z < 5.8$:

$$r(z) = 0.95 \cos\left[\frac{\pi}{2}(z-4)\right] + 2.25. \tag{8.59}$$

(3) Sewing ring, $5.8 \leq z < 6.7$:

$$r_i(z) = \sum_{j=1}^{4} c_{ij}(z - z_i)^{j-1}, \quad z_i \leq z < z_{i+1}, \quad i = 1, \ldots, 7 \tag{8.60}$$

where the values of z_i and coefficients c_{ij} are given in Table 8.1.

(4) Sinus of Valsalva, $6.7 \leq z < 10.0$:

$$r(z) = 0.03 \, (10 - z)^{1.5} \sin\left[\frac{3\pi}{4}(z-6) - \frac{\pi}{2}\right] + 1.54. \tag{8.61}$$

(5) Aortic wall, $10.0 \leq z \leq 25.0$:

$$r(z) = 1.54. \tag{8.62}$$

Table 8.1: Cubic spline coefficients for the sewing ring structure.

i	z_i	c_{i1}	c_{i2}	c_{i3}	c_{i4}
1	5.8000	1.345	-0.41715	0.00000	-43.28503
2	5.9000	1.260	-1.71570	-12.98551	61.42514
3	6.0000	1.020	-2.47005	5.44203	5.41959
4	6.0800	0.860	-1.49527	6.74274	0.39943
5	6.2225	0.785	0.45075	6.91349	0.89021
6	6.3600	0.980	2.40245	7.28071	-42.32405
7	6.4800	1.300	2.32142	-7.95595	12.05447
8	6.7000	1.554	0.57111	0.00000	0.22000

The approach adopted for generating a boundary-fitted grid is that described by Mobley and Stewart [72]. The goal is to evaluate grid point values in the physical plane (z, r) subject to the requirement that the grid lines in the transformed plane (ξ, η) form a conjugate harmonic pair, thereby causing the $\xi - \eta$ contours to be orthogonal. That is, they must satisfy the Cauchy-Riemann equations

$$\xi_z = \eta_r, \tag{8.63}$$

$$\xi_r = -\eta_z. \tag{8.64}$$

In order to control the separation of grid lines in the physical plane, ξ and η are defined as monotonically increasing functions of two variables χ and ζ, respectively. That is,

$$\xi = f(\chi), \quad \eta = g(\zeta). \tag{8.65}$$

Laplace-type equations are formulated for each of the spatial coordinates in the (z, r) physical plane. In general form, the Laplace-type equations are

$$\mathcal{L}(z) = 0, \tag{8.66a}$$

$$\mathcal{L}(r) = 0 \tag{8.66b}$$

where the operator $\mathcal{L}$ is defined as

$$\mathcal{L} \equiv \frac{\partial^2}{\partial \chi^2} + \delta^2 \frac{\partial^2}{\partial \zeta^2} - F \frac{\partial}{\partial \chi} - \delta^2 G \frac{\partial}{\partial \zeta} \tag{8.67}$$

and

$$\delta(\chi, \zeta) = \frac{f'(\chi)}{g'(\zeta)}, \quad F(\chi) = \frac{f'(\chi)}{f''(\chi)}, \quad G(\zeta) = \frac{g'(\zeta)}{g''(\zeta)} \tag{8.68}$$

where $f' = df/d\chi$, $f'' = d^2f/d\chi^2$, etc., and $f' \neq 0$, $g' \neq 0$.

To complete orthogonality the solutions for both z and r are coupled at the domain boundaries by application of the Cauchy-Riemann conditions. A weak constraint is applied so that the final distribution of boundary grid points is not predetermined but is computed through the numerical mapping procedure. For example, in order to solve Eq. (8.66a), z must also satisfy the Neumann condition

$$z_\zeta = -\frac{g'}{f'}r_\chi \tag{8.69}$$

along the top and bottom boundaries, whilst Dirichlet conditions are specified on the other domain boundaries. Similar conditions are applied to the domain boundaries when solving Eq. (8.66b), except that Dirichlet conditions are applied to the top boundary using Eqs. (8.58)–(8.62), to the bottom boundary using the patched hemisphere equation, and a Neumann condition on the remaining walls.

A finite difference method employing central space derivative approximations was developed for solving Eqs. (8.66a) and (8.66b) but the resultant grid was noticeably non-orthogonal, particularly near boundary segments with high surface gradients. Conventional error analysis revealed that the coefficients of higher ordered terms in the vicinity of the steep boundary segments were assuming extremely large values. This problem was partially resolved by switching to either backward or forward differencing (depending upon the orientation of the boundary segment) when computing in these regions.

A further improvement was obtained by implementing a simple procedure described by Davies [18] where orthogonal trajectories are used for the orthogonalisation of a discrete non-orthogonal coordinate system. An iterative scheme is used for the numerical solution of the grid equations. An adequate initial guess is obtained by simple linear interpolation of the boundary values in the radial direction, and equidistant grid lines in the axial direction (see Thalassoudis *et al.* [109]). The "overlapping" of grid lines predominant during early iterations stems largely from the rather poor initial guess and the accompanying complexity of the geometric configuration. Several other initial estimates for z and r involving compression and interpolation of grid lines were attempted, but with little success. After 300 iterations the (ξ, η) coordinate lines were no longer crossing. Convergence, albeit non-orthogonal, was declared after 500 iterations. These values of z and r were then employed as initial values for orthogonalisation using the method of Davies [18]. Excess grid lines in the radial direction were removed and some additional lines were eventually inserted in the axial direction. The final computational grid (120 × 16) is shown in Fig. 8.10.

It should be remembered that the production of orthogonal meshes is by no means an easy task as it depends heavily upon domain geometry. From a practical point of view, it would be more efficient to produce a general coordinate system.

Fig. 8.10: The final computational grid (120×16).

8.7.4 *Finite Difference Formulation*

The transformed governing equations are discretised in the (ξ, η) coordinate system. Central differencing is applied to all derivative terms in Eq. (8.43) and rearranged to yield

$$
\psi_{i,j} = \left\{ 2 \left(\frac{\alpha}{\Delta \xi^2} + \frac{\gamma}{\Delta \eta^2} \right) \right\}^{-1} \times \left[\alpha \frac{(\psi_{i+1,j} + \psi_{i-1,j})}{(\Delta \xi)^2} \right.
$$

$$
- 2\beta \frac{(\psi_{i+1,j+1} - \psi_{i+1,j-1} + \psi_{i-1,j-1} - \psi_{i-1,j+1})}{4(\Delta \xi)(\Delta \eta)}
$$

$$
+ \gamma \frac{(\psi_{i,j+1} + \psi_{i,j-1})}{(\Delta \eta)^2} + \sigma \frac{(\psi_{i,j+1} - \psi_{i,j-1})}{2(\Delta \eta)}
$$

$$
+ \tau \frac{(\psi_{i+1,j} - \psi_{i-1,j})}{2(\Delta \xi)} - \frac{J}{r} \left(\frac{(z_{i+1,j} - z_{i-1,j})}{2(\Delta \xi)} \frac{(\psi_{i,j+1} - \psi_{i,j-1})}{2(\Delta \eta)} \right.
$$

$$
\left. \left. - \frac{(z_{i,j+1} - z_{i,j-1})}{2(\Delta \eta)} \frac{(\psi_{i+1,j} - \psi_{i-1,j})}{2(\Delta \xi)} \right) + J^2 \omega_{i,j} r \right] \tag{8.70}
$$

where the indices (i, j) refer to a computational grid point and $(\Delta\xi)$, $(\Delta\eta)$ are the grid spacings in the (ξ, η) system.

Equation (8.44) is discretised in a similar manner, except for terms such as ω_ξ and ω_η which are treated using upwind differencing for stability purposes. That is, if u and v are the components of velocity in the ξ and η directions, respectively, defined by

$$u = \frac{z_\xi U + r_\xi V}{\sqrt{\gamma}}, \qquad v = -\frac{z_\eta U + r_\eta V}{\sqrt{\alpha}} \tag{8.71}$$

then

$$\omega_\xi \approx \begin{cases} \dfrac{\omega_{i,j} - \omega_{i-1,j}}{(\Delta\xi)}, & \text{if } u \geq 0 \\[2mm] \dfrac{\omega_{i+1,j} - \omega_{i,j}}{(\Delta\xi)}, & \text{if } u < 0 \end{cases} \tag{8.72}$$

and

$$\omega_\eta \approx \begin{cases} \dfrac{\omega_{i,j} - \omega_{i,j-1}}{(\Delta\eta)}, & \text{if } v \geq 0 \\[2mm] \dfrac{\omega_{i,j+1} - \omega_{i,j}}{(\Delta\eta)}, & \text{if } v < 0. \end{cases} \tag{8.73}$$

The one-sided approximations for the vorticity together with the central space approximations for the remaining derivatives are substituted into Eq. (8.44) and rearranged to yield a finite difference equation for $\omega_{i,j}$. Upwind differencing is applied in a similar manner to obtain the k and ϵ finite difference forms from Eqs. (8.55) and (8.56), respectively, but are not reproduced here for brevity.

8.7.5 *Numerical Solution*

The finite difference equations are solved iteratively with relaxation parameters to control stability and convergence. Alternating the direction of the iterative sweeps also enhances stability. The inflow boundary conditions for the ψ, k and ϵ are extended throughout the flow domain as initial conditions. Vorticity values along the wall are at first approximated from the initial stream function values with vorticity values being set to zero elsewhere. Once a numerical solution is obtained, it is used as the initial condition for a higher Reynolds number flow.

The iterative scheme employed is summarised in steps as follows:

(i)   Specify initial values for $\psi^{(1)}, \omega^{(1)}, k^{(1)}$ and $\epsilon^{(1)}$.

(ii) Update the stream function at grid points immediately away from solid boundaries.

(iii) Iterate the stream function at interior grid points with a relaxation factor of 1.75 until convergence. Denote the result $\overline{\psi}^{(n+1)}$.

(iv) Set

$$\psi^{(n+1)} = 0.96\,\overline{\psi}^{(n+1)} + 0.4\,\psi^{(n)}. \qquad (8.74)$$

(v)   Evaluate U, V and G.

(vi) Update the boundary vorticity values.

(vii) Iterate the vorticity at interior grid points with a relaxation factor of 0.9 until convergence. Denote the result $\overline{\omega}^{(n+1)}$.

(viii) Set

$$\omega^{(n+1)} = 0.3\,\overline{\omega}^{(n+1)} + 0.7\,\omega^{(n)}. \qquad (8.75)$$

(ix) Evaluate ν_{eff}, U, V and G.

(x)   Apply 'law of the wall' to grid points immediately away from solid boundaries.

(xi) Evaluate the turbulent kinetic energy at interior points with a relaxation parameter of 0.15. Denote the result $\overline{k}^{(n+1)}$.

(xii) Set

$$k^{(n+1)} = 0.3\,\overline{k}^{(n+1)} + 0.7\,k^{(n)}. \qquad (8.76)$$

(xiii) Apply 'law of the wall'.

(xiv) Evaluate the turbulent kinetic energy dissipation rate at interior points with a relaxation parameter of 0.15. Denote the result $\overline{\epsilon}^{(n+1)}$.

(xv) Set

$$\epsilon^{(n+1)} = 0.3\,\overline{\epsilon}^{(n+1)} + 0.7\,\epsilon^{(n)}. \qquad (8.77)$$

(xvi) Test for global convergence and return to Step (ii) if necessary.

Global convergence is declared when all dependent variables change by less than 1% of their maximum value between iteration steps for at least 50 times. The accuracy of the model numerical solutions was verified by repeating the calculations for $R_e = 1000, 2000$ and 4000 on a 120×31 coordinate system. This was necessary to ensure that a sufficiently dense grid had been chosen for the computations.

Converged numerical solutions for steady turbulent flow through a fully open Starr-Edwards 1260-11A prosthetic heart valve are obtained for Reynolds numbers of

1000 and 6000 corresponding to steady flow rates of 8.133 and 48.796 litres min^{-1}, respectively.

Stream function contours are presented in Fig. 8.11. The solutions show two regions of separated flow. A large separation zone is attached to the downstream side of the ball occluder and a smaller annular zone is attached to the sewing ring and inside the sinus. No separated regions are predicted upstream of the sewing ring nor along the aortic wall. The experimental studies conducted in [22] on flow through a Starr-Edwards 1260-11A caged-ball valve in an aortic-shaped chamber contained separated flow regions attached to the sewing ring (upstream and downstream), the aortic side of the cage struts and distal to the ball occluder. The separation region immediately upstream of the sewing ring was only described qualitatively and appeared to be very small.

Fig. 8.11: Computed mean streamlines for $R_e = 1000$ and 6000.

The computed mean axial velocity profiles are given in Fig. 8.12. At the $z = 30$ mm location, the computed velocity profiles appear to be partly influenced by the narrowing chamber dimensions further upstream. They no longer conform to the fully developed turbulent velocity profiles specified at inflow.

Reverse flow (-38 cm sec^{-1}) present in the sinus separation zone can be clearly seen in the $z = 70$ mm velocity profile. The maximum axial velocity (187 cm sec^{-1}) occurs very close to the upstream surface of the ball occluder. This profile is very similar in shape to the one experimentally measured at an identical location.

Computed wall shear stress profiles are shown in Fig. 8.13 for $R_e = 4000$. Upstream of the sewing ring the wall shear stress remains below 3 Nm^{-2}. The shear stress increases rapidly as the flow encounters the primary valve orifice reaching a maximum of 138.3 Nm^{-2} at the sewing ring tip. The wall shear decreases rapidly downstream of the sewing ring but elevated levels are again observed along the sinus boundary. Along this boundary the wall shear stress attains a negative peak of

Fig. 8.12: Computed mean axial velocity profiles at 30 mm and 70 mm from the inflow boundary for a Starr-Edwards ball valve.

-8.1 Nm^{-2} which is a direct consequence of the velocities in the sinus separation zone. Further downstream, a broad region of elevated wall stress is seen along the aortic boundary corresponding to where the flow impinges upon the wall. The peak level of computed aortic wall shear stress is 30.7 Nm^{-2}. The maximum computed stress along the occluder wall was -63.2 Nm^{-2} for $R_e = 4000$. The level of stress rises and falls rapidly on either side of this peak.

8.7.6 *Remark*

The numerical models discussed in this section using computational fluid mechanics provide detailed quantitative descriptions for the steady flow characteristics of caged-disk and caged-ball prosthetic heart valves. The first model is very simple. It considers laminar flow through a caged-disk valve at several Reynolds numbers. The results indicate the potential effectiveness of such computational studies, but the model is limited in application because of the laminar flow assumption. The second model considers turbulent flow through a caged-ball prosthesis in an aortic-shaped chamber. A boundary-fitted coordinate system is generated and the governing flow equations are conveniently solved on a rectangular transformation plane. The behaviour of the model is generally consistent with experimental observations. Regions of slow moving flow, high fluid stress and elevated wall shear are identified.

(a) Computed aortic wall shear stress (Nm⁻²);

(b) corresponding streamlines; and

(c) occluder wall shear stress (Nm^{-2}) for $R_e = 4000$

Fig. 8.13.

This study presents solutions to a numerical model for steady turbulent flow through a Starr-Edwards caged-ball prosthesis in an aortic-shaped chamber. This valve has an estimated 122,000 worldwide implants, thus underlining the practical importance of this research.

Using computational fluid mechanics techniques to solve complex mathematical models has provided valuable and detailed quantitative data for velocities and stress fields in the vicinity of prosthetic heart valves — a totality currently unattainable through physical simulators.

Ghista [34] pointed out the "unlimited potential" of computer design and graphics for the biomechanical design of prostheses and surgical procedures. Accordingly, the long term application of this form of research is seen both as an effective design tool and to assist cardiac surgeons in the selection of a suitable valve implant for individual patients. The application of mathematical modelling and computational fluid mechanics for the evaluation of the flow characteristics of prosthetic heart valves has and will continue to provide much valuable information.

8.8 Computational Fluid Dynamics Applications to Cardiovascular Health Assessment

8.8.1 *Introduction*

Cardiac dysfunction constitutes common cardiovascular health issues in the society and has been an investigation topic by researchers in the medical imaging community. Diagnostic modalities based on echocardiography, magnetic resonance imaging, chest radiography and computed tomography are common techniques that provide myocardial structural information to diagnose heart defects. However, functional information of cardiac flow, which can in fact be used to support diagnosis of most cardiovascular disease with a myriad of haemodynamics performance indicators, remains unexplored to its full potential. Some of these indicators can characterise the important cardiac functional parameters affecting the cardiovascular abnormalities. With the advancement of computer technology that facilitates high speed computational fluid dynamics, the realisation of a support diagnostic platform for haemodynamics quantification and analysis can be achieved. We will review here the state-of-the-art of medical imaging and high fidelity multi-physics computational approach that presents the haemodynamics parameters, which can discover abnormal trends in the underlying cause of disease such as atherosclerosis and left ventricular dysfunction. This can potentially extend medical knowledge of the defects and their haemodynamic behaviour with an integrated medical image diagnostics and health performance analysis scheme.

The modern medical imaging community in practice is associated with a landscape of non-invasive imaging, image processing and cardiovascular analyses components that constitute the traditional means of assessing cardiovascular system performance. The present-day cardiologists usually diagnose patients with cardiac abnormalities via visual observations of the heart and arterial system from well-established clinical imaging such as echocardiography, magnetic resonance imaging/angiography, chest radiography and computed tomography, and then they provide their expert opinion accordingly. Of recent decades, high fidelity multi-physics computational platforms that solve cardiovascular haemodynamics indictors are studied and may be integrated as a support tool into the existing medical imaging systems to generate a more evident-based patient-specific diagnosis of cardiovascular malfunctions. This type of integrated high performance computational platform in combination with the relevant fluid mechanics, supports medical image diagnosis by quantification of parametric values for decision making, and yields expert opinion for surgical treatment.

Cardiac disease can introduce various cardiovascular defects in the human body, and can be examined using modern imaging methods. We identify briefly some of these possible events. Septal defects, myocardial fibrillations, arrhythmias, heart valve failure and other heart disease can be assessed for degrees of abnormality and evaluation of post-surgical success. Cardiac flow fields that are obtained from clinical ultrasound-based or phase-contrast MR image scan sequences can be analysed, and interesting flow phenomena relating to the operation of cardiac structures such as heart valves can be discovered. Nevertheless, qualitative assessment of the cardiovascular defects does not give a proper evaluation of the underlying cause and effect, as well as comparison of the healthy and abnormal conditions. Appropriate haemodynamics parameters need to be used for quantitative studies. For instance, haemodynamical analysis of the heart structure has been investigated in the vascular network based on localised wall shear stress (WSS) regions, since it is known that the WSS has an implication on atherosclerosis (see Marshall *et al.* [65]). Quantification of blood flow shear stress in various pathologies such as atherosclerotic arteries, the aorta, and in the coronary arteries of the heart can reveal how WSS influence atherosgenesis. We note that this WSS analysis can also be extended to the investigation of artificial cardiac devices such as coronary stents (see LaDisa *et al.* [57]), ventricular assist devices (Wong *et al.* [120]), and heart valves (Haggag *et al.* [36]). Myocardial diseases in particular to their occurrence at the cardiovascular arteries can be examined from the haemodynamics perspective and other various performance indicators can be designed to achieve this. Importantly, the applications derived from the imaging and computational processing can be used to investigate cardiac health in relation to atherosclerosis.

In the context of atherosclerotic arteries, the superficial carotid bifurcation is an ideal target for non-invasive imaging via ultrasound (US), magnetic resonance imaging (MRI) or computed tomographic (CT) imaging. The angiographic techniques such as magnetic resonance angiography, X-ray angiography, etc. are often commonly used in examining atherosclerosis. For example, cineangiograms are used by cardiologists to visually examine the degree of stenosis in carotid bifurcations or coronary arteries. However, it needs to explain the underlying cause of the problem from the flow perspective. It can be demonstrated that the plaque ulceration is related to the existence of high wall shear stress (WSS) at the upstream region of the plaque. On the contrary, some serial MRI study based on carotid artery plaques proved that the regions exposed to low WSS are most prone to develop atherosclerotic plaques. There are also efforts to determine that the investigations of local risk factors such as time-averaged wall shear stress (TAWSS) and oscillatory shear index (OSI) in atherosclerosis need not rely on all the surrogate geometric markers of disturbed flow. The sensitivity of these physiologically relevant parameters to the arterial inlet and outlet boundary conditions may also be investigated by image-based haemodynamics studies (Morbiducci [73]).

From the perspective of the ventricular system, surgical ventricular restoration (SVR) addresses the issue of aneurysms, or large akinetic walls and dilated ventricles. This type of procedure requires knowledge of the shape and size of the left ventricle (LV), as well as information on the intra-ventricular flow dynamics to design useful performance indicators for evaluation of heart pumping efficiency. Computational fluid dynamics come into play when haemodynamics indicators such as the resistance-to-filling and contractility information can be derived based on the computed intra-LV blood flow velocity fields and pressure gradients after transforming the geometry into a computational fluid dynamics (CFD) platform. At the initial stage of the pipe-line, medical imaging of the heart are performed via dedicated scanners and then fed into the image processing pipeline in order to achieve geometrical reconstruction of the cardiac chamber of interest such as the LV. Slices that depict the relevant anatomy are selected based on the region of interest for analysis. Wall boundaries are defined and then used to build a geometrical model that is deformable temporally. Mesh grids are generated and imported into a CFD platform to solve the transient flow fields. Due to the large wall movement, special considerations to the flow simulator needs to be taken care of, for example, the geometric conservation law needs to be used for unsteady flows with moving boundaries. Note that the Navier-Stokes equation is to be solved using a high performance computer that results in the three-dimensional cardiac flow field. Haemodynamics parameters assessing the performance of the LV are then quantified. The information on healthy and normal subjects can be input into a database and statistical quantification based on a range of haemodynamics parameters can give support to cardiologists in terms of diagnosis, expert opinions on treatment, clinical advice, etc.

It may be of interest to experimentally determine the true flow fields which can be used as a more realistic version for analysis of the flow behaviour in the heart. In addition, such information can also be used to validate the numerically computed flow fields. Functional medical imaging modalities are on the rise in terms of speed, accuracy and reliability in measurement of blood flow in the cardiovascular system. On the velocity-encoded (VENC) imaging technology front, phase contrast magnetic resonance imaging (PC-MRI) that is coupled with cardiac flow analysis (Wong *et al.* [120]) gives rise to a system of direct visualisation and analytical processes for describing flow behaviour in cardiac chambers. The use of VENC MR imaging applied onto cardiac imaging of heart chambers can enable a good assessment of haemodynamics that exist in the heart. It may be worthwhile pointing out that another common flow measurement modality is the ultrasound of the heart, which is commonly known as *echocardiography*. We note that just as in the arteries, cardiac flow in the heart chambers generates spiraling motion or vortices and is essential in efficient heart operation and blood circulation. It is of interest now to examine the flow behaviour in the cardiovascular structure in conjunction with any health parameters or indicators.

8.8.2 *Computing Framework for the Assessment of Haemodynamics*

The overall methodology of an experimental validation study is depicted by Fig. 8.14. Numerical studies of haemodynamic characteristics based on computational fluid dynamics (CFD) can be performed based on an anatomically realistic cardiovascular model reconstructed from medical images. The last stage would be the determination of haemodynamics indicators/parameters for cardiac health assessment.

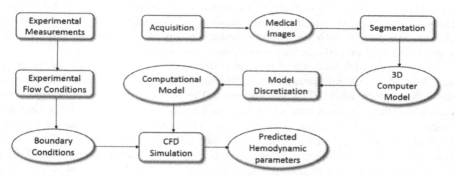

Fig. 8.14: Procedural flow chart for medical imaging and high performance computing of haemodynamics parameters.

This systematic approach performs medical imaging and computational fluid dynamics of cardiovascular structures. Data retrieval and anatomical reconstruction based on medical imaging then generates a geometrical model. Thereafter, the measurement of boundary conditions and using it to set the numerical simulation based on the anatomical reconstructed model can be performed. The predicted data from simulated flows can be visualised, and useful haemodynamics indicators are then extracted for analysis in the final stage.

The usefulness of experimental measurements will be for the validation of numerical simulated results and as a form of support for fluid mechanists to review the accuracy of their simulation platform. In addition, experimental data can also be used to determine the boundary conditions when setting the numerical framework. In the case of such cardiovascular anatomies, velocity-encoded (VENC) phase contrast MRI or ultrasound (US), which are flow measurement techniques, can be used to extract the flow values at the inlets and outlets of the anatomical structures, (see

Groen [33]). The application of the inlet and outlet boundary conditions will affect the numerical accuracy of the local risk factors such as TAWSS and OSI dramatically. Therefore, the imposition of patient-specific inflow and outflow rates based on the US or MRI measurements will be preferred.

The system integration of the components outlined before gives rise to the systematic approach that can be used for assessing the cardiac condition of cardiovascular patients. This forms a generic approach to constructing a system for analysing cardiac defects via examination of the flow variations derived from medical imaging and post-processing techniques.

8.8.3 *Medical Imaging and Anatomical Reconstruction*

In this section, let us review the medical imaging and visualisation application with unique quantification of some of the structural parameters that will lead to a viable diagnostic system for evaluating cardiac related failures or health risk in patients. The ease and speed of preparation are important considerations for cardiologists to quickly gain insights into the malfunctioning regions of the heart and to assist in planning surgical procedures viz., operations or artificial implantations.

The imaging and geometrical reconstruction system is vital because of the technical feasibility of using measured information from medical imaging to determine blood flow information. The key success is the result of implementing high performance computing as a post-processing tool for reconstructed geometry that are produced by images from medical scanners. This renders the encoding of velocity information redundant which reduces scanning and processing time. Medical image reconstruction and high performance computing open up many new opportunities for flow analysis concurrent to the examination of cardiac chambers, septal defects, and heart valve behaviour. It may also be of interest to cardiologists and physiologists to provide information on the structural shape and size of the cardiac chambers in relation to the flow in the heart.

Typically, the initial objective will be to map the cardiovascular structure of the heart chamber or arteries. From medical imaging modalities such as ultrasound, MRI or CT scans of the left ventricle (LV) myocardium, the wall boundary can be extracted and reconstructed to obtain the dynamic LV geometry. In the example presented by Fig. 8.15, medical imaging using the steady-state free precession MRI protocol allow imaging to be registered within the body at localised sections with specific depth and thickness. As explained by Yeo *et al.* [125], the anatomical reconstruction of the LV endocardium into a chamber geometry from the aforementioned steady-state free-precession MR images in the short-axis and long-axis scan orientation can be performed by the triangulation algorithm.

In the above diagram, magnetic resonance imaging is carried out on the neck region of the patient whose carotid artery can be imaged at axial orientation for

Fig. 8.15: Medical image reconstruction of cardiovascular structures.

multiple slices. Segmentation based on the threshold of the blood vessel at various slices is performed in the initial stage. The segmented voxels can be grouped to form a three-dimensional anatomy and a mesh reconstruction based on the contours of these segmented regions is carried out (A). In a similar way, the left ventricle is imaged and ventricular chamber segmentation is performed. Then left surface formation into a geometrical surface structure is enabled to give the anatomical model computationally (B).

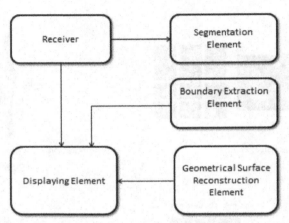

Fig. 8.16: Components of medical imaging and anatomical reconstruction system.

Thus we can examine the elements of a medical imaging and anatomical reconstruction platform that comprise of a generic scanning modality, an image processing and a geometrical boundary definition, and a modelling framework. It is essential to establish a network of sub-components such that each of the components has a role in preparing the anatomical reconstruction. The system layout of this computational reconstruction is presented in Fig. 8.16. The framework serves to produce the computational mesh of the cardiovascular structures, e.g., the cardiac chamber or the artery that are obtained from medical imaging modalities such as MRI, CT, or ultrasound. The core functionality of the system is the visualisation of anatomical structures and identification of their defects. Observation of the structural details generated by the system can assist in identifying stenosis or regions of critical plaque growth in the case of the atherosclerotic arteries, and septal defects in the case of discontinuity in myocardium of the heart which can occur in the atrium or ventricle. It is worthwhile noting that cine-images of the cardiac chamber such as the atrium can be constructed and played using a multimedia tool to observe irregularity in beating and assist in diagnosis of atrial fribillation. In addition, the geometrical mesh of the left ventricle (LV) output by the system can be fed into a simulation modeller and solved for the cardiac flow within using computational fluid dynamics to record useful flow behaviour such as vortices or any other interesting flow patterns within. Useful haemodynamics indicators can also be derived for any of the cardiovascular structures mentioned, which can be used to support medical diagnosis of the cardiac condition.

As explained in Fig. 8.16, medical images can be taken temporally using an imager or scanner. The images are then received by a receiver residing in software on a processor usually in DICOM format and then deciphered and extracted. The segmentation element produces the surface boundaries of an anatomy such as the myocardium. The segmentation element allows a user to assist in the segmentation process.

From the computer architecture perspective, the procedures executed by the system of components can be outlined as follows. After pre-processing, the medical images can be retrieved using a receiver for post-processing at a later stage. Practically, segmentation of the region of interest is semi-automatically determined with external peripherals. In the case of ultrasound or tagged MR imaging, tracking of moving signals encoded onto the medical images can be carried out by a motion estimation component to define the wall boundary. The geometrical structure is then scaled and displayed using a suitable Displaying Element. Optionally, other measures such as the velocity of cardiac wall movement can be computed with a numerical simulation framework using Computational Fluid Dynamics (CFD) and implemented for cardiac flow analysis. Note that the Receiver, Segmentation, Boundary Extraction, Geometrical Surface Reconstruction Elements form the main processing system of this framework. The components within this processing system are utilised for post-processing medical images and to format the geometry of the anatomical structure of interest. Outside of this system is the physical hardware that facilitates the medical scanning, user inputs and display. We will examine each of the components with greater detail in the subsequent sections.

Magnetic resonance imaging is a well-established medical imaging modality and can be used effectively for imaging the heart and arteries with good spatial and temporal resolution. Diagnostic system for detecting cardiac abnormalities and quantifying the degree of cardiovascular defect has proven to be clinically attractive. The information from MRI will have useful future practical benefit for the assessment of cardiac-related failures.

8.8.4 *Reconstructing Surface Mesh and Boundary Conditions*

After extracting the cardiovascular structure by segmentation, the computational file can be saved as an IGES, STL, or STEP in order to be cross-compatible with a range of 3D modelling and meshing programs that can result in different types of mesh configurations as shown in Fig. 8.17. Typically, the meshing procedure begins by application of a simple unstructured tetrahedral mesh all over which produces a single contiguous mesh. However, for easier post-processing of local flow variables, the computational model can be split into smaller sub-regions during the CAD surface and volume generation stage, prior to meshing. While the process of sub-dividing the computational model into smaller regions can be performed within some CFD packages, it is not always an easy task, and therefore it is recommended to be performed in CAD packages that have NURBS functionality.

Fig. 8.17: Mesh configurations of cardiovascular models.

The mesh construction for the stenosed artery is based on an appropriate scale selection as a form of illustration (A) in Fig. 8.17. In this case, the element numbers of the stenosed artery is 696,108. Increased mesh density can be performed at the stenosis region. Note that the density of grid is set to be higher at the region of interest, i.e., the stenotic region (coloured as black). The grid-based mesh of a left ventricular geometrical reconstruction is illustrated in (B). The geometrical mesh of the left ventricle and zoom-in view of the intraventricular grid at its base and apex. The numeric grids of the entire heart can even be reconstructed with variable grid density for the different anatomical structures such as the atria, ventricles, aorta and vena cava, as shown in (C). Images of (B) and (C) in Fig. 8.17 are due to Schenkel *et al.* [96] and Doenst *et al.* [19] respectively.

Generation of a quality mesh is of extreme importance for obtaining reliable computational solutions. A good quality mesh improves numerical stability, and increases the likelihood of attaining a reliable solution. A mesh can be viewed as a number of smaller mesh or grid cells that overlays an entire domain geometry. In general the set of fundamental *Navier-Stokes* equations are applied onto each cell. These equations, which calculate the changes in each cell within the domain, are subsequently solved to yield the corresponding *discrete* values of the flow-field variables such as the velocity, pressure, and temperature.

For the meshing of a cardiovascular structure such as an atherosclerotic artery, surfaces are created and stitched up to create a computational mesh. An initial model with N number of cells is created. The original model is refined by cell adaptation techniques that included refining large volume cells, cells that displayed high velocity/pressure gradients and near wall refinements. This process is repeated twice, with each repeat producing a model with a higher cell count than the previous model. These models are then used in simulation and outputs such as the velocity profiles are compared. Then, the model with the mesh that does not result in a significant difference in flow results compared to the mesh of higher density at its next step, is selected for further runs. A sample model mesh is shown in Fig. 8.17 (A) for the atherosclerotic artery.

In Figs. 8.17 (B) and 8.17 (C), an example of a rectangular mesh for the left ventricular and corresponding computational geometry is demonstrated. The transformation is carried out such that there is a one-to-one correspondence between the rectangular mesh in the computational domain and the curvilinear mesh in the physical domain. The algebraic forms of the governing equations for the multiphase problems are also carried out in the computational domain which has uniform spacing of $\Delta\xi$ and $\Delta\eta$. Computed information is then directly fed back to the physical domain via one-to-one correspondence of grid points. Because of the need to solve the equations in the computational domain, they have to be expressed in terms of the curvilinear coordinates rather than the Cartesian coordinates, which means that they must be transformed from (x, y) to (ξ, η) as new independent variables.

Scan images based on the velocity-encoded MRI modality as shown in Fig. 8.18 are performed for the aorta. VENC MRI enables the direct visualisation of flow without a need to carry out high performance computing of the data for modelling. However, it is unable to predict future flow events such as conditions when a stenosis in an artery worsen or post-surgical treatment is needed in the case of surgical ventricular reconstruction.

The magnitude and velocity-encoded phase images of the aorta are shown in (A) and (B) respectively. The images of cropped phase and intensity encoded velocity of aorta at 16 time phases of a cardiac cycle are illustrated in (C). The intensity

Fig. 8.18: Phase contrast measurement of blood flow in aorta.

of the pixels in the image indicates the magnitude of the velocity component in the specified orientation. Combining two orthogonal velocity-encoded image maps can produce a two-dimensional velocity flow field, which can be used for demonstrating any interesting asymmetric velocity patterns and indicating the presence of helical flow along the channels of the artery. The reconstructed flow fields can be used to validate computerised flow simulations.

For a realistic simulation of cardiac structures, boundary conditions can be well-established and derived based on measurement of cardiac flow profiles obtained from velocity-encoded imaging modalities by the phase contrast MRI scans. The under-standing and derivation of flow properties in such conditions are necessary. The accuracy of the haemodynamics of an isolated patient-specific cardiovascular structure highly depends on the application of the boundary conditions. As such, phase contrast MRI may be used to provide these boundary conditions for the computational model.

In the context of intraventricular flow, the cardiac flow fields of the subject may also be fully measured and compared with the computational model. It may be worthwhile to note that the measured flow patterns can be used as a gauge for checking if the simulation settings are valid. As such, in terms of validating CFD results, imaging modalities such as MRI can be used effectively. Pairs of flow fields generated by CFD simulation and MRI experimentation can be compared quantitatively for the purpose of establishing the computational platform properly prior to further analysis. The two sets of results will have an overall topological structure that agrees well with each other. The flow rates of the haemodynamic structure can be determined and used as boundary conditions for the simulation. It is worthwhile noting that the inflow boundary conditions play an important role in developing reproducible simulations, because the computational model is extremely sensitive to these imposed conditions.

8.8.5 *Haemodynamics Performance Indicators*

Velocity vector plot and axial profiles arising from computational haemodynamics can be used to assess influence of the flow in the model. The flow patterns can be assessed by velocity vector as well as flow profile plots. From the velocity vector plots and axial velocity profiles, we can examine regions of flow separations, accelerating and decelerating flows, as well as their transient magnitudes. However, in order to assess the health conditions of the atherosclerosis from the velocity information, a more precise indicator should be deployed to examine the disease more appropriately.

Let us now discuss some parameters commonly used to assess blood flow. Certain haemodynamics parameters require a clear definition prior to performing simulation of the blood in the circulatory system comprising the arteries and heart.

Resistance of blood to deformation under shear stress causes viscosity. Practically, we can describe its internal resistance to flow as a form of fluid *friction*. Unlike many fluids which behave like Newtonian fluids with a constant viscosity, blood, which consists of plasma, blood cells and other material carried throughout the blood stream, tends to behave like non-Newtonian due to the quantity of particles within the plasma. In fact, the viscosity now changes with the shear rate of the flow. Realistically, under normal conditions it is not viable to ignore the non-Newtonian behaviour of the fluid.

In a fluid flow that is incompressible and under a steady flow field, the shear rate of strain of a fluid element is defined as the rate of decrease of the angle formed by two mutually perpendicular lines on the element. As such, the shear strain rate $\dot{\gamma}$ is proportional to the rate of decrease of axial velocity v_z along the arterial radius:

$$\dot{\gamma} = -\frac{dv_z}{dr}. \tag{8.78}$$

The rate of change in velocity along the radial section from wall to centre of the vessel, previously defined as the shear strain rate, is proportional to the wall shear stress. The mechanical characteristics of the flow can be described by the simple law, which defines a linear relationship between the shear stress τ and shear strain rate with the viscosity μ as its gradient. Therefore, the equation for wall shear stress is given by:

$$\tau = \mu\dot{\gamma}. \tag{8.79}$$

Blood flows along curved arteries with complex flow dynamics such as variable axial velocities along the radial section. In the case of carotid arteries, flow velocity varies in magnitude at regions close to the other wall and near the bifurcation. The complex phenomenon is responsible for the time-dependent oscillatory wall shear stress distribution throughout the blood vessel.

Three commonly adopted flow indicators to evaluate the total shear stress exerted on the wall throughout a cardiac cycle are the time-averaged wall shear stress (TAWSS), the oscillatory shear index (OSI), and the relative residence time (RRT), which can be presented as shown in the following equations:

$$\text{TAWSS} = \frac{1}{T}\int_0^T |\tau_w| dt, \tag{8.80}$$

$$\text{OSI} = \frac{1}{2}\left[1 - \frac{|\frac{1}{T}\int_0^T \tau_w dt|}{\frac{1}{T}\int_0^T |\tau_w| dt}\right], \tag{8.81}$$

$$\text{RRT} = \frac{1}{(1 - 2 \times OSI) \times TAWSS}, \tag{8.82}$$

where T is a cardiac cycle period and τ_w is the instantaneous wall shear stress. The atheroprotective effect of the endothelial cells is influenced by these haemodynamic indicators. Low TAWSS values (lower than 0.4 Pa), high OSI (higher than 0.5) and high RRT (higher than 10 m^2/N) are known to promote an atherogenic endothelial phenotype (see Morbiducci et al. [73]), while abnormally high TAWSS (higher than 40 Pa) values can cause direct endothelial injury and increase the risk of thrombosis (see Malek et al. [62]).

The same computational approach can also be applied in the context of evaluating the performance of heart pumping such as the efficiency in contraction and dilation throughout the cardiac cycle. Computational fluid dynamic modelling can also provide the quantitative and qualitative evaluation of the surgical ventricular reconstruction procedure. This can facilitate preoperative modelling for patients to achieve optimised post-SVR flow performance. By knowledge of these parameters, we can then predict and assess the heart pumping efficiency based on information provided by CFD.

The fluid dynamical parameters of such nature include the cardiac cycle time T_0, the Womersley number W_o, and the Reynolds Number Re, as well as the exchange transfusion that is denoted by M, which represents the remaining fractional blood volume in the ventricle after an arbitrary number of cardiac cycles n (see Doenst et al. [19]). We have

$$\text{M} = (1 - \text{EF})^n. \tag{8.83}$$

As can be observed from the computed data, the exchange transfusion M, shows consistently higher value for the patient as compared to the healthy normal subject, which indicates that more of the blood volume remained in the ventricle at the end of every cycle. On the contrary, the intraventricular condition of the patient is such that it has a much less efficient wash-out. Furthermore, the Womersley number W_o, is demonstrated to be moderately lower for such a patient as compared to the healthy subject.

It is worthwhile mentioning here that traditional diagnosis of left ventricular (LV) dysfunction is based on cineangiograms, ultrasound monitored or MRI scans of LV wall motion. Observation of the improvement in wall motion leading to a higher ejection-fraction after administration of myocardial vasodilators can determine if coronary bypass surgery is necessary. The following flow related quantification parameters can also be determined: (a) end-diastolic volume (in cardiovascular physiology, end-diastolic volume (EDV) is the volume of blood in the right and/or left ventricle at end load or filling in diastole or the amount of blood in the ventricles just before systole) and end diastolic pressure (EDP), (b) stroke volume (SV) which is the volume of blood pumped from one ventricle of the heart with each beat, and stroke work (SW). The stroke work is defined as the measurement of how much energy is converted to work by the heart and thus is the product of the mean aortic pressure (the average blood pressure throughout the cardiac cycle) and the stroke volume (the amount of blood pushed into the aorta with each beat of the heart (end-diastolic - end-systolic)). Thus, the systolic pressure, stroke volume, cardiac output, stroke work, and cardiac work represents the ability of the ventricle to function as a pump (known as the Frank-Starling relation). The stroke work is the left ventricular work/beat. The right ventricle is only one-seventh of the left ventricle because the pulmonary vascular resistance is less than the systematic vascular resistance. The stroke work can also be found through mechanical efficiency and arterial inefficiency. Stroke Work (SW) = Stroke Volume (SV) * Mean Aortic Pressure against which the blood is ejected by the left ventricle. The other remaining parameters are (c) LV resistance-to-filling and contractility, (d) ejection fraction (EF) which is the fraction of outbound blood pumped from the heart with each heartbeat, and is commonly measured by echocardiogram and serves as a general measure of a person's cardiac function, i.e., EF=SW/EDV, (e) maximal rate of change of pressure-normalised stress, and (f) regional function in terms of change of the endocardial surface area (S) for end diastole (ED) to end systole (ES).

The LV systolic properties are also useful parameters to look at when determining the LV performance and systolic function indices. Another important parameter to look at is the global shape of the LV, which is characterised by a sphericity index defined as the ratio of the short axis to the long axis. As such, these performance parameters, in addition to the fluid mechanical parameters determined by the computer simulation, can be used together to support the health of the cardiac chamber.

8.8.6 *Concluding Remarks*

In general, cardiac medical imaging has been widely utilised for imaging patients with various cardiac conditions, such as diseases of the aorta, aneurysm, and human hearts with atrial and ventricular septal aneurysm or defect. However, traditionally, only anatomical information of the cardiovascular compartments or structures is used for diagnosis, until the development of phase contrast MRI that is able to provide cardiac flow quantification. On the numerical modelling front, CFD is able to provide simulation of various scenarios or conditions such as pre- and post-surgical treatment whereby the actual measurement could not take place if it has not even occurred. We summarise here the clinical applications of medical imaging in the cardiovascular system for which the following medical conditions are assessed (see English and Moore [21]): (i) Pericardial disease; (ii) Congenital heart disease; (iii) Aortic arch heart disease; (iv) Acquired heart disease; (v) Cardiac transplantation; (vi) Atrial and ventricular septal defects; (vii) Valve regurgitation; (viii) Aneurysms; and (ix) Coarctation of the aorta. The quantification of flow characteristics within the heart and arteries provides vital information to cardiologists interested in a range of problems from the haemodynamic of blood to biological phenomena in the heart. Flow inefficiencies such as whirlpools or swirling structures, as well as turbulence can also be examined, and their existence can ultimately be linked to the operation of a number of cardiac structures influencing flow dynamics in the heart. This has potential applications for identifying risks for heart failures, stroke and for plaque vulnerability.

It is important to note that vessel or heart chamber imaging and visualisation is not sufficient as a stand-alone information provider of the cardiac health assessment. The important health performance indicators discussed before can be coupled along with the qualitative information as a potential diagnostic system for the assessment of heart disease and as a practical tool for physiological analysis. For example, in addition to the evaluation of atherosclerosis and heart pumping performance, haemodynamics performance indicators also have the potential applications for identifying risks after heart valve implant as well as determining the degree of atrial or ventricular septal defects. The simulation may also be used to examine the growth effect of cardiac tissue into the mesh of surgically implanted device from the fluid mechanical perspective, and haemodynamics indicators such as vorticity can be designed to assess its health performance (see Wong *et al.* [121], [122]).

Quantification of cardiac health that uses an integrated image processing and high performance computing approach may be used on a patient-specific basis without the need for additional diagnostics or in-vivo procedures, and thereby making it attractive to future clinical practice. Healthcare organisations will benefit by using the simulated haemodynamics performance indicators for evaluation of cardiac health and using the flow fields as an assessment for analysis of flow phenomena in cardiac pathology. Manufacturers of medical imaging machines may incorporate such computer visualisation techniques into their imaging systems. The image data can be transferred to dedicated computer laboratories with high performance computing facilities to generate the appropriate haemodynamics health indicators. Medical organisations (from the cardiac discipline) can then assess the health of human heart and arteries so that appropriate medical action can be taken.

Being at the cutting edge of medical science, the recent developments in the fields of medical imaging modalities have given new dimensions to our understanding of the human cardiovascular system. In fact, this concept can be integrated into the present medical image diagnostics by well-established computational modelling for determination of appropriate haemodynamics indicators. Cardiac flow analysis contributes the framework behind the operation of future flow simulation or mapping systems using high performance computing. This may lead to a new perspective, which will be obtained using insights from the field of medical imaging and computer modelling. Further development in the emerging field of medical imaging and visualisation, and high performance computing will significantly contribute towards health benefit.

Medical imaging in combination with high performance computing will ultimately impact on future medical image diagnostics. The computational flow indicators and visualisation to couple with anatomical details will be a significant advancement for technological development in the next era. The introduction of new diagnostic procedures for the evaluation of heart defects, as well as the capability of the research to identify and quantify flow phenomena through it will potentially result in clinical information and give cardiologists a leading edge in saving patients.

Chapter 9

TISSUE ENGINEERING

9.1 Introduction

Tissue Engineering (TE) is a new interdisciplinary field dealing with the principles of engineering and life sciences toward the development of biological substitutes that restore, maintain, and improve tissue function or a whole organ (Bronzino [13]). It represents a combination of cellular and molecular biology on one hand and material engineering on the other. Currently, TE is emerging as an invaluable field of study and is one of the most promising fields for the next century. The ultimate goal is to develop powerful new therapies for biological substitutes that will successfully restore the structural and functional disorders of human and animal system. TE is an incredibly progressive field of medicine that may in future, help with the shortage of life-saving organs available through donation for transplantation.

Fig. 9.1: Possible locations of different cartilages in human body (*Source:* [1]).

Available standard treatments for organ or tissue loss include transplantations of the same organ or organs from healthy donors, surgically reconstructed with synthetic implants in most of the cases, and the use of mechanical devices such as kidney dialysers, ventricular assist device (VAD) etc. (Webstar [116]). Of course, these treatments can save lives and decrease morbidity. However, each has its significant drawbacks and most are associated with substantially decreased quality of patient's life. Although these procedures (treatments) temporarily save countless lives, unfortunately these techniques are not perfect.

Tissue engineering procedures offer a new hope in modern medical science. By using patient's own cells to produce bioengineered tissues and organs that function like their natural counterparts, TE may offer a better solution for the repair and reconstruction of damaged tissue and organs. At present, the only successfully engineered tissue that has been made available to people is skin. Major success in this direction came in 1984 when Robert Langer, a chemical engineer at MIT and Joseph Vacanti, a medical surgeon at Harvard, first time jointly came up with an idea of duplication of tissues by developing a *biodegradable scaffold* on which skin cells could be grown using *fibroblasts* [56]. After rigorous and extensive experiments with different techniques they discovered that collagen and fibroblasts can arrange themselves in a compact dense form under specific physical conditions that is exactly similar to the inner layer of the skin. Seeding *keratinocytes* on this layer leads to the formation of upper layer of the skin. This is how the *artificial skin* was developed by TE methods. The important point of this artificial skin is that it has no immunogenic cells, therefore question of rejection does not arise when implemented on patient's body. Artificial skin are now made available commercially in large quantities. The next major step towards the development of artificial organ using tissue engineering approach is the cartilage. It is believed that the cartilage has relatively simple composition and structure.

9.2 Cartilage Tissue Engineering

9.2.1 *Introduction*

We shall consider first human joints system. A number of human joints viz., a *synovial joint*, a *knee joint*, a *hip joint* etc. play important roles in tissue engineering of human joints. A *synovial joint* may be described as a load-carrying system consisting of two mating bones with normal and/or shear motions. The bone ends, which are usually globular in appearance, are layered with a soft sponge-like material, called *articular cartilage*. The space between these

Fig. 9.2: Damaged knee.

cartilaginous extremities of the bones, known as joint cavity, is filled with a shear-dependent fluid, known as *synovial fluid.*

In a *knee joint* (Fig. 9.2), the femur (upper bone) and the tibia (lower bone), the two major bones of the leg, undergo relative motion during walking. In the *hip joint*, which is a ball-and-socket joint, the head of the femur forms the ball of the joint and the acetabulum of the pelvis the socket. The hip joint is regarded as separated by spherical surfaces but the knee joint as separated by cylindrical surfaces. The load support area is larger for a hip joint than for a knee joint. The structure of the knee joint is such that it can support a load nearly four times the body weight.

The behaviour of a *synovial joint* is mainly governed by the characteristics of the articular cartilage and synovial fluid, named after "synovia" (egg white). The synovial is a clear yellowish dialyzate of blood plasma with added hyaluronic acid, containing complex proteins. Its chief constitute is water. The synovial fluid exhibits a non-Newtonian pseudo-plastic behaviour. When bones approach one another, water and other low molecular weight molecules pass through the pores of the cartilage thus increasing the concentration of these molecules in the synovial fluid and increasing its viscosity. In a diseased joint, the cartilage surface may become rough and cracked and synovial fluid may lose its non-Newtonian behaviour. The pores of a cartilage may increase in size.

Cartilage is a remarkably complex biocomposite material that exhibits an outstanding comprehensive stiffness, toughness, strength, resiliency and shock absorption. There are three different types of cartilage: *articular cartilage, elastic cartilage* and *fibro cartilage.* We are particularly interested in articular cartilage only. Articular cartilage is a relatively simple tissue, covers the end of the long bones in joints and enables the bones to move smoothly over one another. Functional characteristics of the articular cartilage involve load bearing (compressive forces) and generating a smooth surface that allows low friction movement in and around the joints. Damage to the articulating surface can occur due to compression of heavy load or when angular and/or shear forces are applied on it.

As mentioned earlier, articular cartilage is a living tissue but of very limited capacity to repair itself. In fact, any damage to the articulating surface (due to injury, disease, and/or genetic disorders) tends to spread allowing the bones to rub directly against each other and resulting in increasing joint pain and loss of joint movement. It can also lead to osteoarthritis (a slow degeneration of articular cartilage) (see Fig. 9.2). It is the most common form of arthritis in Australia, about 10% of the total population. In United States alone, more than one million people annually suffer from osteoarthritis, a degenerative joint disease caused by damage to articular cartilage [64]. Damage to the articulating surface can occur due to compression of heavy load or when angular and/or shear forces are applied on it. With the prevalence of cartilage-related disease (e.g., Osteoarthritis), injuries and cost of treatment will be on the rise. An overview of a healthy and damaged knee is given in Fig. 9.2.

The ability of cartilage to withstand enormous compressive loads (many times the body weight) without being crushed, is attributable to the multi-phasic nature of the tissue. Mature articular cartilage contains approximately 5% of its volume as cells (also known as the growth cells or *chondrocytes*), and 95% as *extracellular matrix* (ECM). From an engineering standpoint, ECM is a porous, viscoelastic material consisting of three principal phases (a phase represents all of the chemical compositions with similar physical properties): 1) the solid phase constitutes about 30% of the total weight, and contains collagen (60%), proteoglycans (PG) (25%) and wide range of matrix proteins (15%), 2) the fluid phase, known as the synovial fluid as explained before, which constitutes on average about 70% of the total weight, for normal tissues, and 3) an ion phase, consisting of many ionic species of dissolved electrolytes (e.g., Na^+, Ca^{++}, Cl^-). These constitute less than 1% of the net weight. The biofluid mechanics of cartilage growth and degradation relies on this multi-phasic nature of the tissue.

Existing techniques to repair damages of articular cartilage either with fibrous tissue or with the use of transplants such as autografts/allografts have very limited success. Tissue engineering approaches to cartilage repair have potentially important advantages over the modern generation of metal-and-plastic joint prostheses. In this arrangement, ECM concentration in cell-polymer constructs is designed to gradually increase and form neo-tissue while scaffold concentration decreases with time. However, the role of ECM in maintaining and modulating cell function is not fully understood although recent reports have described the role of receptor molecules such as integrins in regulating chondrocytes.

9.2.2 *Collagen*

The collagen fibrils are densely packed polymer strands with a high resistance against fluid flow. They are the dominant structural components of the solid matrix and helps in retaining the shape of the cartilage when compressed. Collagens are basically made of three polypeptide chains arranged as triple helix long its length. There are two different types of collagens, known as *fibrillar* and *non-fibrillar* collagens. Depending upon structural and functional characteristics collagen has been classified as 12 different categories. Hyaline cartilage (articular cartilage) mainly consists of type-II and type-XI fibrillar collagens where the percentage of type-II collagen is much higher.

Fig. 9.3: Proteoglycan & collagen distribution in articular cartilage (*Source*: [92]).

The architecture of articular cartilage's connective tissues is very complicated. Collagen fibrils in the superficial-most region (known as the tangential zone) of the cartilage layer are densely packed and oriented parallel to the articular surface. This collagenous membrane has a relatively low PG content and a lower permeability to fluid flow, which is important in providing for a barrier of high resistance against fluid flow when cartilage is compressed. In the middle or transitional zone of the cartilage, the collagen fibers are larger and are either randomly or radially orientated. In the deepest zones, i.e., the zone nearest to the calcified cartilage and sub-chondral bone interface the collagen fibers are larger and form bundles that are oriented perpendicular to the calcified/bony interface. The way how the collagen and proteoglycan/glycosaminoglycans molecules are arranged inside the articular cartilage, is shown in Fig. 9.3.

9.2.3 *Proteoglycan*

Proteoglycan (PG) are macromolecules consisting of a protein core to which are attached short chains of negatively charged glycosaminoglycans (GAG). The primary GAGs associated with PGs in cartilage are chondroitin 4-sulfate, chondroitin 6-sulfate and keratan sulfate (Fig. 9.4). It is available in all connective tissues but its structure varies with the type of tissue. Hyaline cartilage contains mainly aggregating proteoglycans.

Fig 9.4: A GAG molecule attached to the core protein in proteoglycan

Each PG-associated negative charge on the polymer chain (known as the fixed charge, FC), requires a mobile counter-ion (e.g., Na^+) dissolved within the interstitial fluid in the cartilage gel to maintain its electro-neutrality. This effect gives rise to an imbalance of mobile ions across the gel interface. The excess of mobile ions colligatively yields a swelling pressure, known as the osmotic pressure, while the swelling pressure associated with FCs is known as the *Donnan pressure*. Changes in this internal swelling pressure, arising from altered ion concentrations of the external bath, or changes in the fixed charges result in changes in tissue dimensions and hydration. However, this swelling pressure is balanced by tensile forces generated in the collagen network. This effect is due to the presence of covalent cross-links within the polymer matrix. An increase in cartilage tissue hydration, governed by the density and the nature of FCs on the PGs as well as the density of the mobile counter ions in the interstitial fluid are the earliest signs of articular cartilage degeneration during osteoarthritis.

9.2.4 *Chondrocyte*

Chondrocytes are mature cells found in the cartilage. They make up the cellular matrix of cartilage, performing a number of functions within the tissue, including production and maintenance of the cartilaginous matrix and facilitating the exchange of fluids through the gelatinous layers. Fully mature chondrocytes tend to be round, and they may cluster together in small groups within the network of the cartilage (Fig. 9.5).

The progenitors of these cells arise in the bone marrow, in a form of stem cell. Stem cells are capable of differentiating into several different types of cell, depending on the need. When they differentiate into cartilage cells, they start out as *chondroblasts*, actively producing secretions of *chondrin*, the primary substance in cartilage, to build and repair the tissue. Once a chondroblast becomes totally surrounded, it is a mature chondrocyte. These cells can be found in small gaps within the cartilage known as *lacunae*.

Chondrocytes are not capable of cell division. They can produce secretions to support and repair the cartilage matrix, and as explained above, they facilitate the exchange of materials between the cartilage and the surrounding material. Depending on what type of cartilage a cell is in, it may have a slightly different composition; elastic, hyaline, and fibrocartilage are all unique, designed to meet various needs within the body.

Fig. 9.5: Light micrograph of hyaline cartilage showing its chondrocytes and organelles, lacunae and matrix (*Source*: [64]).

9.3 Mathematical Model of the ECM in Tissue Engineering Study

Mathematical models are essential for understanding systems (like cartilage) with many interacting components (e.g. the collagen network, proteoglycans, interstitial fluid and ionic solutions present in cartilage), since the complexity of the interactions taking place overwhelms the ability of the human intuition to make accurate predictions of their behaviour. Comparisons of the models' quantitative predictions with experimental results allow us either to confirm that the biological hypotheses regarding these interactions (which are encoded in the model) are correct or, equally usefully, identify that some of them are in error. Once validated, the model will ultimately help design treatments related to cartilage loss and osteoarthritis by predicting this outcome for proposed interventions, and focusing efforts on those which may appear most promising.

A relatively strong negative correlation has been observed between GAG molecule deposition within ECM and its synthesis rate in the study of native and engineered cartilage. This result is consistent with that reported from *in vitro* modulation of chondrocyte metabolism by extracellular proteoglycan. Based on this product inhibition mechanism, a mathematical formulation for GAG deposition was described by Maroudas [64] and adapted for use in cartilage tissue engineering by Wilson *et al.* [118] for a characterisation of ECM mass or concentration:

$$\frac{d(ECM)}{dt} = k_{ECM}\{(ECM)_{SS} - (ECM)\} \tag{9.1}$$

where k_{ECM} is an ECM synthesis rate constant and the steady-state (ss) condition occurs at $t \to \infty$ and $d(ECM)/dt = 0$. Thus ECM mass or concentration as a function of time follows the form:

$$(ECM)_t = (ECM)_{SS}\left[1 - e^{-k_{ECM}t}\right]. \tag{9.2}$$

Degradation of scaffold mass or concentration is also assumed to follow a first order decay equation:

$$\frac{d(scaffold)}{dt} = -k_s(scaffold) \tag{9.3}$$

where k_s is the scaffold decay rate constant. Again, this implies at time-dependent characterisation as:

$$(scaffold)_t = (scaffold)_{initial}e^{-k_s t}. \tag{9.4}$$

By combining the mathematical formulations (9.2) and (9.4), along with the cellular mass content, a total construct mass model is proposed and experimentally validated as a function of culture time. In the above approach, dependencies between chondrocyte function and ECM molecule production rates as well as the effect of local biomechanical or biochemical factors are not explicitly taken into account.

It is well established that growth factors and cytokines modulate cell function and cartilage development. The effects depend on specific combinations of factors, time sequences of their application, and interactions with the mechanical environment (hydrodynamic shear and compression). It has been observed experimentally that growth factors play an important role in matrix production and inhibition, but there is limited knowledge about the functional mechanisms associated with these effects. This leads to the objective which is to explore the influence of growth factors within engineered cartilage by modelling hypothetical characteristics of ECM synthesis.

In previous studies, predictive models which examined deterministic and stochastic mechanisms of engineered cartilage composition are presented (see Saha, Mazumdar and Kohles [94]). These works examined cellular proliferation effects, a randomised influence of growth factors on synthesis rate, a deterministic approach with exponential synthesis terms, and a randomised influence of growth factors limited by a carrying capacity parameter with an initial growth factor state. Here, dynamic compositional models are again formulated albeit with concomitant growth and decay rate terms. These modelling iterations investigate possible influences in the modulation of ECM synthesis which can be manipulated to enhance the regenerative potential in cartilage therapies. Therefore, the specific aims of this study are to: a) develop a deterministic mathematical model based on published experimental results that describe the accumulation of matrix molecules using distinct rate terms; b) quantify a 'growth factor effect' on matrix molecule deposition through a stochastic influence on the synthesis rate term. Both of these modelling approaches are based on current cell-polymer constructs developed for tissue engineered cartilage.

In this section, an account of the equilibrium for electro-chemical properties of polyethylene gylcol gel (PEG) containing negatively charged chondrotin sulfate (ChS) strands, a material which closely mimics articular cartilage in electro-mechano-chemical properties, immersed in water with several different ions: H^+, Na^+, Cl^- is demonstrated. This material closely mimics the electro-chemical properties of articular cartilage. We view this mixture as a multi-component material, including solvent particles, polymer and particles of several ionic species. Each polymer chain is assumed to be made up of two types of *monomers* (monomers are the basic, non-repeating units within the chain): the uncharged units (i.e., PEG) and the charged ones (i.e., ChS). Only the charged units participate in the binding reactions, carry a double negative charge and are denoted as M^{2-}. The positively charged ions in the solvent are Hydrogen (H^+) and Sodium (Na^+). The negatively charged ions could include Hydronium (OH^-) and Chloride (Cl^-). Because the negatively charged ions are assumed to be not involved in any binding reactions with the gel, acting only as counterions to positive charges, these ions are identified by the name Chloride. The binding reactions of the positively charged ions with the monomers are:

$$(a) \ M^{2-} + H^+ \underset{k_{-h}}{\overset{k_h}{\rightleftharpoons}} MH^-, \quad (b) \ M^{2-} + Na^+ \underset{k_{-n}}{\overset{k_n}{\rightleftharpoons}} MNa^-, \quad (c) \ MH^- + H^+ \underset{k_{-h2}}{\overset{k_{h2}}{\rightleftharpoons}} MH_2,$$

$$(d) \ MNa^- + Na^+ \underset{k_{-n2}}{\overset{k_{n2}}{\rightleftharpoons}} MNa_2, \quad (e) \ MH^- + Na^+ \underset{k_{-hn}}{\overset{k_{hn}}{\rightleftharpoons}} MHNa. \tag{9.5}$$

We assume that all the binding sites/charge sites are identical and the binding affinities for the different ions are different. The species M^{2-}, MH^-, MNa^-, MH_2, MNa_2 and $MHNa$ are different monomer species of the charged type, all of which move with the same polymer velocity. The ion species are freely diffusible, but because they are ions, their movement is restricted by the requirement to maintain electroneutrality.

Suppose we have some volume V of a mixture comprised of k types of particles each with particle density (number of particles per unit volume) n_j, and particle volumes ν_j, $j = 1, \cdots, k$. Let us denote the quantities with subscript 1 related to PEG monomer species, subscript 2 related to ChS monomer species and subscript 3 related to solvent molecule. The subscripts p and s, denote polymer and solvent phase, respectively. The volume fraction for each of the polymer species (θ_1, θ_2) and the solvent $(\theta_3 = \theta_s)$ are $\theta_i = \nu_i n_i$ $(i = 1, 2, 3)$ respectively, and $\theta_p = \theta_1 + \theta_2$. Suppose the ChS monomers constitute a fraction α of the total number of monomers, n_p, (i.e., $n_2 = \alpha n_p$, $n_1 = (1 - \alpha)n_p$). Assuming that the ionic species do not contribute significantly to the volume (i.e., we take $\nu_j = 0$, $j = 4, \cdots, k$), conservation of total volume implies $\theta_1 + \theta_2 + \theta_3 = 1$. Further, in subsequent calculations we make an assumption that the particle density of the ions is insignificantly small compared to the particle density of polymer and solvent, i.e., $\sum_{j\geq 4} n_j \ll n_s + n_p$. Hence, $\phi_s = \frac{n_s}{n_p+n_s}$, $\phi_p = \frac{n_p}{n_p+n_s}$, $\phi_j = \frac{n_j}{\sum_{i\neq 1,2} n_i}$ for $j \geq 4$ (assuming that the ions are dissolved in the solvent), are the polymer, solvent, and ion species per total solvent particle fractions, respectively.

9.3.1 *Chemical Potential*

This section outlines the calculation of the free energy and the chemical potential for this multi-species material. As before, suppose there are k different kinds of particles, with particle volumes ν_i and particle numbers $n_i, i = 1, \ldots, k$. To calculate the chemical potentials, μ_j, we use the Gibb's free energy

$$G = -k_B T S + U + PV, \qquad (9.6)$$

where U is internal energy, S is entropy, T is temperature, k_B is Boltzmann's constant, P is pressure, and $V = \sum_j \nu_j n_j$ is the total volume of the system. However, we assume that the volume occupied by species $j = 4, \cdots, k$ (which are the ions dissolved in the solvent) is small compared to that of the monomer plus solvent volumes $\nu_p n_p + \nu_s n_s$ and so take $V = \nu_p n_p + \nu_s n_s$. The chemical potential are given by

$$\mu_j = \frac{\partial G}{\partial n_j} = -k_B T \frac{\partial S}{\partial n_j} + \frac{\partial U}{\partial n_j} + \nu_j P = \mu_j^S + \mu_j^I + \nu_j P, \qquad (9.7)$$

where μ_j^S, μ_j^I are the contribution due to entropy and internal energy respectively. In the next two sections these two contributions to the chemical potential are outlined.

9.3.2 *Entropic Contributions to Chemical Potentials* (μ_j^S)

The entropy of the system is defined as

$$S = \sum n_i \omega_i, \qquad (9.8)$$

where ω_i is the entropy per particle for the ith particle. Using standard counting arguments given by Flory [24], for single-molecule species, we have

$$\omega_j = -\ln(\phi_j) \qquad j \geq 3. \qquad (9.9)$$

The PEG and ChS chains exhibit permanent cross-link bonds (i.e., covalent bonds), and we calculate the per-particle entropy of these two species using the Doi's rubber elasticity theory [20]

$$\omega_i = -\frac{3k_i}{2N_i}\left[(\phi_i)^{-2/3} - 1\right], \qquad i = 1, 2, \qquad (9.10)$$

where $i = 1, 2$ corresponds to PEG and ChS polymer species respectively. The corresponding particle fractions are $\phi_1 = (1 - \alpha)\phi_p$ and $\phi_2 = \alpha\phi_p$. k_i and N_i are the fractions representing the number of cross-linked monomers per total number of monomers in one chain, and the number of monomers per chain, respectively. Finally, the expressions for the entropic chemical potentials can be calculated (see Keener *et al.* [53, 54] and Sircar *et al.* [100]).

9.3.3 *Internal Energy Contributions to Chemical Potentials* (μ_j^I)

The internal energy consists of two contributions, long range electrostatic interactions and short range (nearest neighbour) interactions. The long range electrostatic interactions have energy

$$U_e = \sum_j z_j n_j \Phi_e, \tag{9.11}$$

where z_i is the charge on the ith ionic species ($z_1 = z_2 \equiv z_p$ is the average charge per monomer), and Φ_e is the electric potential.

To calculate the short range interaction energy for the polymer and solvent, we assume that each of the n_T ($= n_p + n_s$) particles have z neighbouring interaction sites (called the coordination number). Of the total of n_p monomers, $k_1 n_1$ and $k_2 n_2$ of them are the cross-linked PEG and ChS monomers, respectively. Since the cross-linked particles are pair-wise connected, we treat each of these pairs as a single species.

The different species with their pairwise interactions (i.e., (a) the cross-linked species, x, (b) the uncross-linked species, u, and (c) the solvent particles, s) are shown in Fig. 9.6. Double solid lines denote cross-linking, single solid lines denote interactions within a polymer chain while the other types of interactions are denoted by dashed lines. The nearest neighbour interactions are assumed identical for both the PEG and ChS monomer species. The following assumptions are made while calculating the per-particle interaction energy, U: (1) all cross-links are assumed identical, (2) the cross-links are covalent permanent bonds, (3) the polymers have long chains and hence the end-effects are neglected, and (4) different polymer species have the same interaction energies if they are of the same type (e.g., uncross-linked PEG and uncross-linked ChS monomers have the same interaction energy). The cross-linked particle pair and the uncross-linked monomer have $2z$-6 and z-2 free interaction sites based on their position in the middle of the polymer chain, respectively (Fig. 9.6(a) and Fig. 9.6(b)). The solvent particles have z free interaction sites (Fig. 9.6(c)).

The interaction energies for crosslinked, uncrosslinked polymer particles, and solvent particles are calculated using standard mean-field arguments. For example, the interaction energies for cross-linked particles, F_x (Fig. 9.6a), are calculated by noting that there are either $\frac{n_1}{2}$ pairs of only PEG monomers, $\frac{n_2}{2}$ pairs of only ChS monomers or $\frac{n_1+n_2}{2}$ pairs of both PEG and ChS monomers, respectively. The corresponding probabilities of forming cross-links are k_1^2, k_2^2 and $k_1 k_2$. For each of these pairs, two neighbouring sites are occupied by the covalent cross-link, four neighbouring inter-

Fig. 9.6: Different species with their pairwise interactions.

actions are the monomer-monomer interactions within the polymer strands, and the remaining $z - 6$ interactions are distributed to monomer-monomer, with probability $\frac{n_p}{n_T}$ and monomer-solvent, with probability $\frac{n_s}{n_T}$. Hence,

$$\frac{F_x}{k_B T_0} = \left(\frac{n_1}{2}k_1^2 + \frac{n_2}{2}k_2^2 + \frac{n_1 + n_2}{2}k_1 k_2\right)(2\epsilon_{xx} + (2z - 6)E_{us} + 4\epsilon_{pp}), \tag{9.12}$$

where $E_{us} = \epsilon_{uu}\frac{n_p}{n_T} + \epsilon_{us}\frac{n_s}{n_T}$. Similar arguments are made to calculate the interaction energies for uncrosslinked particles (F_u) and solvent particles (F_s). The total per particle interaction energy is

$$\begin{aligned}
U &= \frac{1}{2(n_m + n_s)}(F_x + F_u + F_s) \\
&= k_B T_0 \left[\chi \phi_p \phi_s + \mu_0^s \phi_s + \mu_0^p \phi_p + \frac{z}{2}\epsilon_{us}\right], \tag{9.13}
\end{aligned}$$

where the coefficients, χ, μ_0^p, μ_0^s, are the Flory interaction parameter and the chemical potentials of pure polymer and solvent species respectively (see Maroudas *et al.* [64]). These are functions of the particle fraction of the ChS strands α, the interaction energy associated with the uncross-linked monomer-monomer interaction ϵ_1, solvent-solvent interaction ϵ_2, covalent cross-links interaction ϵ_3 and monomer-monomer bond within a polymer strand ϵ_4. The factor $\frac{1}{2}$ in the per-particle interaction energy in Eq. (9.13) is to correct for double counting. The corresponding contributions to chemical potentials μ_j^I are calculated using Eq. (9.7).

9.3.4 *Interface Conditions*

Under no-loading conditions, there is a free moving-edge to the gel on one side (inside the gel) $\theta_p = \theta_p^-$, $\theta_s = \theta_s^-$, and on the other side (outside the gel) $\theta_p^+ = 0$, $\theta_s^+ = 1$, we have the interface conditions,

$$\frac{1}{\nu_p}\mu_p^- \mathbf{n} = 0, \tag{9.14}$$

for the polymer, and

$$\frac{1}{\nu_s}(\mu_s^+ - \mu_s^-)\mathbf{n} = 0,$$ (9.15)

for the solvent. Here, $\mathbf{n}$ is the normal to the free surface. The interface conditions are derived using the standard variational arguments to minimize the rate of work, which constitutes the viscous rate of energy dissipated within the polymer and the solvent, energy dissipation rate due to the drag between solvent and polymer and between solvent and ion species particles as well as the rate of work required against the chemical potential, μ_j. These interface conditions contain the contribution to the swelling pressure via the entropy, the Donnan potential and the net osmolarity (or the swelling pressure coming from the difference between the concentrations of ions dissolved in the gel and those dissolved in the bath, outside the gel). The details of this derivation can be seen in Keener *et al.* [53, 54] and Sircar *et al.* [100].

9.3.5 *Ionised-species Chemistry*

Let us use the law of mass action to describe the chemical reaction. Under the assumption of fast chemistry and diffusion, the forward and backward reaction rates in Eq. (9.5) are balanced. Those relations provide us with quasi-equilibrium concentration of the monomer species (e.g., $m = [\text{M}^{2-}], y = [\text{MH}^-], v = [\text{MNa}^-], w = [\text{MH}_2], x = [\text{MNa}_2]$ and $q = [\text{MHNa}]$). Similarly, the quasi-equilibrium concentration of the ion species ($n = [\text{Na}^+], h = [\text{H}^+]$ and $c_l = [\text{Cl}^-]$) reduces to

$$C = C_b e^{-z_C \Psi_e},$$ (9.16)

where C is the chemical concentration of the ions inside the gel ($C = h, n, c_l$), and the z_i is the ion-valence ($z_n = z_h = 1$ and $z_{c_l} = -1$, respectively) and the subscript 'b' denotes the corresponding bath concentrations (i.e., concentrations outside the gel). The electrostatic potential, Ψ_e, is determined by the electro-neutrality constraint inside the gel, namely,

$$(n + h - c_l)\theta_s + z_p m_T = 0,$$ (9.17)

where z_p is the average charge per monomer. This charge depends on the residual charge of the unbound and the bound ChS monomers, z_p, given by

$$z_p m_T = -(2m + y + v).$$ (9.18)

From Eqs. (9.17), (9.18), we can calculate the Donnan potential, $z_p \Psi_e$. This potential is zero outside the gel (since there is no polymer in that region).

9.4 Experimental Methods

In vitro experiments provide the most intuitive mechanism for cartilage swelling-deswelling kinetics. Recent experiments have confirmed that changes in external ion concentration and pH affect not only the mechanical properties of cartilage, but also the biosynthetic capabilities of chondrocytes. The experimental outcomes of a number of studies made previously share a common feature: collagen accumulation in engineered cartilage tend toward a constant value (steady state) with respect to time. A similar type of result has been found for GAG molecule accumulation in engineered cartilage constructs.

According to the results on DNA content of engineered cartilage constructs obtained by Saha and Mazumdar [95] (Fig. 9.7), there is an increasing trend of cell population growth that means the cell population has no control, i.e., it keeps on increasing unboundedly, but this is not the case in nature. Hence, overall experimental observations suggest that ECM deposition in cell-polymer construct of engineered cartilage heading towards a steady state. However, there is no concrete evidence that chondrocytes population does the same.

Fig 9.7: DNA concentration *Source:* [95]

The experimental validation of the model, described previously, can be performed by running a subset of *in vitro* experiments in tandem with the computational experiments as described below. Construct ionic gels by co-polymerising methacrylated chondroitin sulfate and poly(ethylene glycol) dimethacrylate in deionised water (distilled water). PEG is chosen since it can be used as a neutral, cross-linked, 3-D scaffold, a system that promotes the deposition of proteoglycans and collagen molecules and emulates the mechanical strength, load bearing capabilities and resilience of cartilage tissue. The ChS component serves two purposes: to mimic the biochemical environment of cartilage (since it is the main component of proteoglycans) and to introduce fixed negative charges into the network. These polymer gels are then allowed to reach equilibrium in deionised water and then placed into appropriate solution (e.g., with variable pH, salt concentration, etc.) to reach a new equilibrium.

To quantify the local configuration of the polymer gels, the counter-ion concentrations and the osmolarity inside these gels, the following in-vitro experiments can be performed and the results can be utilised to validate the outcome of numerical simulations: (i) Firstly, vary the particle fraction of PEG:ChS in a salt-free, neutral bath ($H_b = 10^{-7}$ M), where α, the particle fraction of ChS, is: 0.1, 0.2, 0.3, 0.4, 0.5 and 1. This experiment is to be performed to determine the role of fixed charges. Since the ChS monomer volume and the density are constant, increasing the particle fraction of ChS will directly correspond to increasing its weight fraction. (ii) Secondly, vary the bath concentration of the monovalent salt within a range of concentrations, 0.1-1 M, and (iii) Thirdly, repeat experiments (i) and (ii) in a salt-free bath and variable pH for neutral hydrogels (i.e., $\alpha = 0$) as well as for ionic gels (i.e., $\alpha > 0$). The pH is then varied in the range 3-8 to ensure that the model accurately captures the local variables (i.e., the volume fractions and the osmolarity), within the physiological range. Further, the above mentioned three types of experiments are repeated for gels with different cross-link fractions. Gels of varying cross-link density are achieved by changing the total monomer concentration in solution prior to polymerisation (e.g., 10 wt %, 20 wt %, and 30 wt % gel). Polymer volume fraction is determined from equilibrium swollen volume and dry polymer volume measurements and compared to the numerically determined value. The initial volume of the dry polymer, V_i, can be determined from the initial mass, m_i, using a weighted density of PEG and ChS in the form

$$V_i = 0.1 \left(\frac{m_i \cdot \%PEG}{\rho_{PEG}} + \frac{m_i \cdot \%ChS}{\rho_{ChS}} \right). \tag{9.19}$$

The mass of the gel at equilibrium can be measured and used to determine the swollen volume, $V_{swollen}$. It is assumed that all polymer is incorporated in the gel, which is confirmed for chondroitin sulfate, and therefore a change in mass is solely due to the a change in the water content. Hence, for polymers with a 10 wt % dry weight in solvent, we have

$$V_{swollen} = 0.1 \left(\frac{m_i \cdot \%PEG}{\rho_{PEG}} + \frac{m_i \cdot \%ChS}{\rho_{ChS}} \right) + \frac{m_{swollen} - 0.1 m_i}{\rho_{solvent}}. \tag{9.20}$$

The initial mass, m_i, is the mass of the gel immediately following polymerisation and before being placed in a solvent. The swollen mass, $m_{swollen}$, is the mass of the gel at equilibrium. The volume fraction of the polymer is calculated as the volume of the dry polymer (i.e., before swelling) divided by the volume of the swollen gel, $V_i/V_{swollen}$. The kinetic chains associated with methacrylate group represent less than one percent of the total gel volume and therefore need not be considered in the volume calculations.

9.4.1 *Comparison of Model Outcomes with Experiments: Parameter Estimation*

The values of the parameters used in numerical calculations are listed in Table 9.1. The constants in the model are the monomer volumes, ν_1, ν_2, the coordination number of the PEG-ChS polymer lattice, z, and the nearest neighbour interaction energies, ϵ_i (see Sec. 9.3.3). The undetermined constants are the binding affinities of the polymer gel with sodium and hydrogen.

The monomer volumes can be found from the density and the molecular weight information ($\nu_i = M_i/(\rho_i * N_A)$). The cross-linked PEG-ChS matrix has a 3-D configuration, which suggests that we choose the coordination number, $z = 6$, which mimics a 3-D cubic lattice. The standard free energies, $k_B T_0 \mu_p^0$ and $k_B T_0 \mu_s^0$ and the interaction energies, ϵ_i are found from the Hildebrand solubility data, δ_i (see Barton [7]). The standard free energy is the energy of all the interactions between the molecule and its neighbours in a pure state that have to be disrupted to remove the molecule from the pure state.

Table 9.1: Parameters common to all the numerical results. The reference temperature for the solubility parameters is fixed at $T_0 = 298$ K.

	PEG ($i = 1$)	ChS ($i = 2$)	Units
Density (ρ_i)	1.07	1.001	g/mL
Molecular weight (M_i)	4600	48700	g/mol
Repeat unit per chain (N_i)	102	86	–
Cross-link fraction (k_i)	0.25, 0.5, 0.75	0.25, 0.35, 0.45	–
Hildebrand solubility (δ_i)	17.39	5.19	MPa$^{1/2}$
Hildebrand solubility for water (δ_w)		48.07	MPa$^{1/2}$

9.5 Effect of Structural and Environmental Fluctuations on Equilibrium ECM Configuration

Articular cartilage is composed of highly negatively charged aggrecan aggegates dispersed in a crosslinked collagen network. This chemical make-up gives rise to an unusually high concentration of mobile-counter ions (e.g., Na+) in the interstitial fluid of the tissue. As a result, the cartilage tissue experiences a complex set of electromechanical cues (signals) arising from fixed negative charges (aggrecan), and mobile-counter ions, in equilibrium. Understanding and recapitulating the electromechanical cues in a tissue engineering approach may be a key to bioengineering functionally competent cartilage.

9.5.1 *Effects of Changes in Ionisation, Cross-links*

Figure 9.8(a), (b), (c) presents the equilibrium volume fraction, corresponding Donnan potential and the net-osmolarity versus the particle fraction of ChS, α, respectively. The various curves represent gels with different cross-link fractions, k_1, k_2, as given in Eq. (9.10). It can be shown that the average charge per monomer (or the average ionisation) increases in proportion to the particle fraction, α. The gel is dissolved in salt-free, neutral water. Any far field boundary effects are neglected in numerical simulations, by assuming infinite bath conditions. The negative charges on the gel cause water to dissociate into hydrogen (H^+) and hydronium (OH^-) ions. The free H^+ ions bind with the monomers.

At low ionisation (or low particle fraction values, α), the physically relevant solutions swell as a function of increasing α. This is due to the increased swelling effect in response to the Donnan potential which can be seen at the trend of the volume fraction and the Donnan potential in the range, $0 \leq \alpha \leq 0.9$. The physically relevant solutions are the largest and smallest ones when there are three solutions, because the intermediate solution is unstable. Another trend is that the gel deswells with increasing α, at high cross-link fraction and particle fraction values (the curve denoted by CL_3 in the range $\alpha > 0.9$, Fig. 9.9(a)). This observation is supported by experimental results that a highly cross-linked network provides a barrier against the

Fig. 9.8: (a) Equilibrium polymer volume-fraction, (b) Donnan potential, and (c) Net-Osmolarity in neutral salt-free, bath conditions vs. α, the particle fraction of chondrotin sulfate component for gel solutions with different cross-link fractions.

Donnan potential and deswells at higher ionisation levels [94].

9.5.2 *Effects of Changes in the Bath Salt Concentration*

Figure 9.9(a),(b),(c) highlights the corresponding equilibrium volume fraction, Donnan potential and the net-osmolarity versus the bath salt concentration, $[NaCl]_b$, for polymer gels immersed in a neutral water solution, respectively. These gels have variable cross-link fractions but a fixed ChS particle fraction, $\alpha = 0.1$. Usual features emerge at high concentrations ($[NaCl]_b \geq 10^{-3}$ M) and low concentrations ($[NaCl]_b \leq 10^{-8}$ M) of salt. At higher salt concentrations, the gel deswells and the deswelling is more prominent for specimen with greater cross-link fraction. This is readily understood since at higher salt concentrations, more dissolved Na^+ ions are furnished which can bind with the negatively charged gel. As a result, the primary ingredient of swelling pressure, i.e., the Donnan potential, is negligible (Fig. 9.9(b)). High cross-link density within the gel resists the swelling pressure which further enhances the deswelled state of the gel. At lower concentrations of salt, there are hardly any Na^+ (or H^+) ions readily available to bind with the gel. The Donnan potential is nearly constant and this feature leads to a fixed equilibrium state of the gel ($\theta_p \approx 0.65$).

For intermediate salt concentrations (10^{-8} M $< [NaCl]_b < 10^{-3}$ M), a new swelling-deswelling feature emerges, namely, there is a swelling, either gradual or through a phase transition, as the salt concentration in the bath increases. The explanation for these swelling features is that there is a complicated interplay between ionisation which promotes swelling and increases in bath concentration of salt which promotes deswelling (see Sircar *et al.* [100]).

9.5.3 *Effects of Changes in the Bath pH*

Two sets of numerical experiments are performed to predict the equilibrium volume fraction of the charged gel. In the first experiment, cross-link fraction of the polymer matrix is fixed at $k_1 = k_2 = 0.25$, while the ChS particle fraction is varied. Overall, it is found that increasing the bath concentration of hydrogen leads to deswelling. The reason is that a decrease in pH leads to an increase in the dissolved cations which readily bind with the gel, thereby decreasing the gel ionisation and the swelling effect due to the Donnan potential. In this case, swelling is predominantly driven by the ionisation which is induced by the Donnan potential. The pressure created by the Donnan potential is about an order of magnitude greater than potential induced osmotic pressure, which is apparent if we compare the order of magnitude of the curves in Fig. 9.10(b) versus Fig. 9.10(c).

Another set of numerical simulations can as well be performed to determine the equilibrium solutions of gels having a fixed ChS particle fraction but a variable cross-link fraction. It can be found that gels progressively preferred a deswelled state in

Fig. 9.9: (a) Equilibrium polymer volume-fraction, (b) Donnan potential, and (c) Net-Osmolarity, vs. NaCl-concentration (in mol/lt.) in the bath. Different curves represent samples with a fixed ChS particle fraction, $\alpha = 0.1$, and variable cross-link fractions.

the order of increasing cross-link fraction values, consistent with our observations in the previous section. In summary, for the values of parameters examined, the gel swells predominantly due to effects of increasing Donnan potential and deswells due to the combination of increasing bath concentration of cations and increasing covalent cross-links within the polymer matrix.

Fig. 9.10: (a) Equilibrium polymer volume-fraction, (b) Donnan potential, and (c) Net-Osmolarity vs. hydrogen bath concentration H_b, for a gel with fixed cross-link fraction $k_1 = k_2 = 0.25$ and variable composition of ChS.

9.5.4 *Concluding Remarks*

In this chapter a quantitative description of the makeup of the cartilage tissues is explained. The study provides a multi-phase, multi-species model to shed light on the swelling / de-swelling mechanism for ionic gels with covalent cross-links; as well as a detailed description of the experimental methods for these biogels. These polymeric biogels mimic the electro-chemical environment of articular cartilage. Finally, the effect of structural and environmental fluctuations on the equilibrium tissue configuration have been adequately demonstrated.

This chapter deals with determining the counter-ion concentration and local osmolarity in dynamically loaded gels, evaluating the effect of the environmental fluctuations on the cartilage cell response and the tissue development. A synchronous development of biomechanical models and numerical schemes are discussed. Some interesting studies in this area have previously been made by Villanueva *et al.* [114] on the experimental side and by Maroudas [64] on the numerical aspect. Indeed, in future, the tissue engineers are expected to answer the fundamental question, *how important are electro-mechano-chemical cues in guiding cellular response and cartilage tissue growth?* Information gleaned from models and experiments will help understand the pathway to cartilage degeneration, repair strategies, maintaining healthy joints via delayed tissue ageing and developing superior mechano-chemically resistant biogels as joint replacements.

Chapter 10

CELLULAR ENGINEERING

10.1 Introduction

Cellular Engineering (CE) can be defined as the application of the principles and methods of engineering to the problems of cell and molecular biology of both basic and applied nature. As biomedical engineering has shifted from the tissue level to the cellular level, cellular engineering has emerged as a new area and is intrinsically connected to the new field of tissue engineering (see Nerem [76]). In the previous chapter, we indicated that the principal aim of Tissue Engineering (TE) is repairing, regenerating, or replacing damaged tissue. The process involves the culture of living human cells in biodegradable scaffold and subsequently allowing them to develop into a three-dimensional tissue (Langer and Vacanti [56]). A series of biological principles, characteristics and techniques are required to understand the entire process. First step is assortment, separation, and culturing of primary cells (progenitor or stem cells). Second step is to promote their differentiations to precise phenotypes. Third step involves seeding the cells for the desired phenotype on bio-degradable scaffolds of biodegradable polymers. The scaffolds typically take the form of meshes, foams, or fleeces, and once seeded with cells, the cell-polymer construct can be cultured *in vitro* in a bioreactor or static culture or *in vivo* in an animal model. Synthetic scaffolds degrade passively in culture, as the cells proliferate and assemble an extracellular matrix (ECM), which in turn replaces the scaffold.

Some useful terminology in this context is explained. *Morphogenesis* which means genetically controlled and systematic development of spatial distribution of cells that gives rise to tissue and organs in early embryonic development. Preliminary study that

213

involves genetic control of cell growth, differentiation and morphogenesis is known as *developmental biology*. Tissue engineering procedure aims at replicating these complex cellular dynamics in the laboratory to construct replacement tissue and/or organs for regenerative medicine (Reddi [83]).

Before exploring the tissue engineering process which has been described in the previous chapter, we need to understand the process of developmental biology. In this section we explore these concepts in two phases: (i) epigenetic mechanism (Fig. 10.1), and (ii) morphogenesis. The term *epigenesis* refers to the fundamental problem of embryology, the perennial mystery of birth of an animal from a single fertilised egg cell (known as *zygote*), be it a frog or mouse or man. The process is further divided into two categories, *Transcription* and *Translation* (Goodwin [32] and Rosen [89]).

Transcription is the process of synthesis of different size classes of RNA (ribonucleic acid) viz. ribosomal RNA (rRNA), transfer RNA (tRNA), messenger RNA (mRNA), heterogeneous nuclear RNA (hRNA) etc. from deoxyribonucleic acid (DNA). The information contained in gene, i.e., DNA is transferred to mRNA during transcription.

Translation is the process in which mRNA chain is decoded leading to formation of protein, i.e., conversion of nucleotide text into amino-acid text.

The foundation of theoretical biology in terms of elementary mathematical expression is due to Mendel. As described by E. Schrödinger [98] in his book *What is Life*, genes are aperiodic which means the same information sites are not repeated at regular intervals. In his version, an aperiodic system contains greater amount of non-redundant information than an equivalent periodic system. Some researchers analysed the genetic information transfer on the basis of purely physical phenomenon. The evolution of form and structure in the biological system is due to morphogenesis that involves the active motion of cells due to mechnochemical process simulated by molecular signalling (Volkenshtein [115]). In his epoch-making work *On the Chemical Basis of Morphogenesis*, Turing [111] investigated the problem of stability towards diffusion to interpret some biological phenomena determining the ontogenetic development. Reshevsky [84] build up a few mathematical descriptions on the basis of diffusion and chemical gradient inside a membrane bound region to analyse the criteria of cell division. Further studies are made on the generation of inhomogeneities in the biological systems from the point of view of instabilities.

Fig. 10.1: Epigenetics mechanism (transcription & translation) (*Source: National Library of Medicine (NLM)*).

In order to use science and engineering principles in biological systems, let us first express it in a mathematical language. From the point of view of systems biology, the cellular system can be constructed as a dynamical model by means of a differential equation and/or a system of differential equations as follows:

$$\frac{d\vec{X}}{dt} = F(\vec{X}, T) \tag{10.1}$$

where

$$\vec{X} = (x_1(T), x_2(T), \ldots, x_n(T))^t \in \Re \tag{10.2}$$

and

$$F(\vec{X}, T) = (F_1(X, T), F_2(X, T), \ldots, F_n(X, T))^t. \tag{10.3}$$

Here $x_i(T)$'s are physical variables, i.e., the concentrations of the reactants in the cell at a particular instant T, involving small molecules, metabolites, DNA, different size classes of RNA, protein etc. Therefore, the state of a cell at any instant can be denoted by a single point in the Euclidean space, called in this context as a state-space. The functions $F_i(\vec{X}, T)$'s are in general non-linear in nature arising due to reactions between $x_i(T)$'s taking place within the cell.

As stated earlier, *transcription* is the process of synthesis of ribonucleic acid (RNA) from strands of deoxyribonucleic acid (DNA), situated in the chromosomes in the nucleus of a cell and translation is the process of synthesis of protein chains on ribosome in cytoplasm by mRNA messenger carrying molecule. The two main physical laws (a) induction and (b) inhibition are actually responsible for the controlled synthesis of these biomolecules (RNA and protein).

Induction means the activation of the particular genes by specific enzymes, called activators.

Inhibition means the repression of the genes by another class of enzymes on metabolites, called *inhibitor* or *repressor*. In biological systems, induction plays the role of positive feedback process whereas inhibition plays the role of negative feedback process.

10.2 System Dynamical Model Building for the Epigenetic Mechanism

Most of the dynamical control systems are based on Jacob and Monod operon concepts. (Jacob and Monod [45]). There are three main regions of DNA, (i) regulator genes, (ii) operator genes, and (iii) structural genes. A special regulator genes produce repressor molecules and those repressor molecules inhibit the operator gene which controls the transfer of information from several structural genes responsible for protein synthesis.

Translating a biological phenomenon into a mathematical framework is not an easy task. The following example illustrates how one can develop a simple comprehensive mathematical model that can explain the protein synthesis involving both transcription and translation. As an example, consider the mathematical framework which is based on the following assumptions (see Tapaswi and Saha [107]):

(i) hRNA (heterogeneous nuclear RNA) is directly proportional to the activation of mRNA genes and inversely proportional to the synthesis of repressor molecules,

(ii) rRNA synthesis is directly proportional to the activation of rRNA genes and inversely proportional to the concentration of tRNA molecules,

(iii) tRNA synthesis is directly proportional to the tRNA genes and inversely proportional to the tRNA molecules itself,

(iv) mRNA synthesis is directly proportional to the amount of hRNA,

(v) aminoacylation of tRNA is directly proportional to the available tRNA,

(vi) polysome formation is directly proportional to the available amount of rRNA and mRNA,

(vii) protein synthesis is directly proportional to the available amount of aminoacyl tRNA and polysomes,

(viii) synthesis of repressor molecules are directly proportional to the amount of protein, and

(ix) each of the above components decays at a rate which is proportional to its own amount of accumulation.

The schematic representation of the epigenetic control mechanism during embryonic development is shown in Fig. 10.2. G_1, G_2, and G_3 are the respective genetic loci from which rRNA, hRNA, and tRNA are transcribed. The minus sign denotes the negative feedback. Amino acid is derived from the intracellular pool, and RAT denotes aminoacyl tRNA, i.e., tRNA charged with aminoacid (see Saha and Tapaswi [93]).

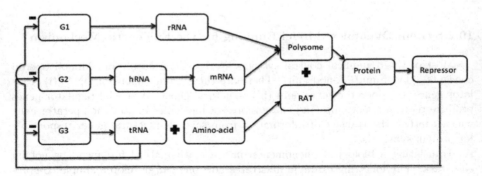

Fig. 10.2: Protein synthesis.

The epigenetic mechanism using the above scheme can be expressed as follows:

$$\text{rRNA genes} \xrightarrow{\alpha_1/(1+h_1 x_4^{m_1})} x_1^f,$$

$$x_1^f \xrightarrow{\beta_1} \text{nucleotides},$$

$$\text{hRNA genes} \xrightarrow{\alpha_2/(1+h_2 x_8^{m_2})} x_2,$$

$$x_2 \xrightarrow{\beta_2} x_3^f,$$

$$x_3^f \xrightarrow{\beta_3} \text{nucleotides},$$

$$\text{tRNA genes} \xrightarrow{\alpha_4/(1+h_4 x_4^{m_3})} x_4,$$

$$x_4 + (\text{amino acid}) \xrightarrow{\alpha_5} x_5,$$

$$x_4 \xrightarrow{\beta_4} \text{nucleotides},$$

$$x_1^f + x_3^f \xrightarrow{\alpha_6} x_6 (= x_1^p + x_3^p \text{ complex}),$$

$$x_6 \xrightarrow{\beta_6} \text{nucleotides} + \text{protein},$$

$$x_5 + x_6 \xrightarrow{\alpha_7} x_7 + x_4 + x_6,$$

$$x_7 \xrightarrow{\alpha_8} x_8,$$

$$x_7 \xrightarrow{\beta_7} \text{amino acid},$$

$$x_8 \xrightarrow{\beta_8} \text{amino acid}. \tag{A}$$

Here, x_1^f and x_3^f are the number of free rRNA and free mRNA molecules respectively, x_1^p and x_3^p are the number of polysomal rRNA and polysomal mRNA molecules respectively. Further, $x_i (i = 2, 4, 5, 6, 7, 8)$ denote respectively hRNA, tRNA, aminoacyl-tRNA, polysome, (a rRNA-mRNA complex), amino acid units of protein and amino acid portion of repressor molecules per cell at any time T. Here, $\alpha_i (i = 1, \dots, 8)$, $\beta_i (i = 1, \dots, 8)$, $h_i (i = 1, 2, 4)$ and $m_i (i = 1, 2, 4)$ are rate constants of synthesis, degradation, inhibition and stoichiometry (Hill's coefficients of inhibition of the respective reactions) respectively of the ith components.

All the above reactions are catalysed by specific enzymes. We further assume that the number of amino acid molecules is very large and constantly replenished from outside extracellular capsule so that this may be taken as constant. Hence, the epigenetic mechanism during embryogenesis obeying the above reaction scheme can be represented by the following set of eight simultaneous nonlinear differential equations:

$$\frac{dx_1^f}{dT} = \frac{\alpha_1}{1 + h_1 x_4^{m_1}} - \beta_1 x_1^f - \alpha_6 x_1^f x_3^f, \tag{10.4a}$$

$$\frac{dx_2}{dT} = \frac{\alpha_2}{1 + h_2 x_8^{m_2}} - \beta_2 x_2, \tag{10.4b}$$

$$\frac{dx_3^f}{dT} = \beta_2 x_2 - \beta_3 x_3^f - \alpha_6 x_1^f x_3^f, \tag{10.4c}$$

$$\frac{dx_4}{dT} = \frac{\alpha_4}{1 + h_4 x_4^{m_3}} - \beta_4 x_4 + \beta_5 x_5 - \alpha_5 x_4 + \alpha_7 x_5 x_6, \tag{10.4d}$$

$$\frac{dx_5}{dT} = \alpha_5 x_4 - \beta_5 x_5 - \alpha_7 x_5 x_6, \tag{10.4e}$$

$$\frac{dx_6}{dT} = \alpha_6 x_1^f x_3^f - \beta_6 x_6, \tag{10.4f}$$

$$\frac{dx_7}{dT} = \alpha_7 x_5 x_6 - \beta_7 x_7, \quad (\beta_7 = \alpha_8 + \bar{\beta}_7), \tag{10.4g}$$

$$\frac{dx_8}{dT} = \alpha_8 x_7 - \beta_8 x_8. \tag{10.4h}$$

The above eight-dimensional nonlinear dynamical system is quite complicated and seems to be mathematically intractable. Hence, the above system is to be simplified to some extent, retaining the main features of the system not significantly altered.

In order to do so, we first consider $x_1 = x_1^f + x_1^p$ and $x_3 = x_3^f + x_3^p$ for each cell at any time T. The molecule x_6 contains one molecule of x_1^p and one molecule of x_3^p, and the rate change of x_1^p and x_3^p is the same as that of x_6 as shown in Saha and Tapaswi [93]. If we assume $\beta_1 = \beta_3 = \beta_6$, then Eqs. (10.4a) and (10.4c) can be replaced by

$$\frac{dx_1}{dT} = \frac{\alpha_1}{1 + h_1 x_4^{m_1}} - \beta_1 x_1, \tag{10.5a}$$

$$\frac{dx_3}{dT} = \beta_2 x_2 - \beta_3 x_3, \tag{10.5b}$$

and Eq. (10.4f) can be written as

$$\frac{dx_6}{dT} = \alpha_6 (x_1 - x_1^p)(x_3 - x_3^p) - \beta_6 x_6. \tag{10.6}$$

Assume $x_1 \gg x_1^p$ and $x_3 \gg x_3^p$ at all stages of the development. This indicates that the total rRNA and mRNA is much in excess than required to form polysome (x_6) molecules in the system. Hence equation for x_6 can be written as

$$\frac{dx_6}{dT} = \alpha_6 x_1 x_3 - \beta_6 x_6. \tag{10.7}$$

We now modify the 11th reaction step in the epigenetic scheme (A) by assuming that x_4 liberated in the reaction is instantaneously converted into x_5. This assumption is validated by the fact that each tRNA molecule carrying the specific amino acid remains still highly activated after discharging the amino acid to the elongating protein chain site and thus immediately catches another specific amino acid from the large pool to become aminiacylated again. Hence the 11th reaction step in (A) can be simplified as:

$$x_5 + x_6 + \text{amino acid} \xrightarrow{\alpha_7} x_5 + x_6 + x_7.$$

Equations (10.4d) and (10.4e) can be simplified as:

$$\frac{dx_4}{dT} = \frac{\alpha_4}{1 + h_4 x_4^{m_3}} + \beta_5 x_5 - (\alpha_5 + \beta_4)x_4, \tag{10.8}$$

and

$$\frac{dx_5}{dT} = \alpha_5 x_4 - \beta_5 x_5, \tag{10.9}$$

respectively. Hence replacing Eqs. (10.4a), (10.4c), (10.4d), (10.4e), (10.4f) by Eqs. (10.5a), (10.5b), (10.7), (10.8), (10.9) the simplified eight-dimensional mathematical model governing the epigenetic control system during embryogenesis can be rewritten as:

$$\frac{dx_1}{dT} = \frac{\alpha_1}{1 + h_1 x_4^{m_1}} - \beta_1 x_1, \tag{10.10a}$$

$$\frac{dx_2}{dT} = \frac{\alpha_2}{1 + h_2 x_8^{m_2}} - \beta_2 x_2, \tag{10.10b}$$

$$\frac{dx_3}{dT} = \beta_2 x_2 - \beta_3 x_3, \tag{10.10c}$$

$$\frac{dx_4}{dT} = \frac{\alpha_4}{1 + h_4 x_4^{m_3}} + \beta_5 x_5 - (\alpha_5 + \beta_4)x_4, \tag{10.10d}$$

$$\frac{dx_5}{dT} = \alpha_5 x_4 - \beta_5 x_5, \tag{10.10e}$$

$$\frac{dx_6}{dT} = \alpha_6 x_1 x_3 - \beta_6 x_6, \tag{10.10f}$$

$$\frac{dx_7}{dT} = \alpha_7 x_5 x_6 - \beta_7 x_7, \tag{10.10g}$$

$$\frac{dx_8}{dT} = \alpha_8 x_7 - \beta_8 x_8. \tag{10.10h}$$

It is now easier to view this complex dynamics of cellular kinetics from a cellular engineering point of view. To obtain the desired outcomes we may have to control the parameters in the system.

Analytical solution to the above system of differential equations is quite a formidable task. From mathematical point of view the system in Eqs. (10.10) can be reduced to five-dimensional problem using the concept of singular perturbation technique. Using Centre-manifold theory (Carr [15]), the five-dimensional system can be reduced further to a two-dimensional system. To get an understanding of the qualitative nature of the system, such as, Hopf bifurcation and limit cycle system may be transformed into Poincaré normal form. It was observed that the system in Eqs. (10.10) processes a globally stable limit cycle around the local unstable equilibrium point to some parametric conditions. However, the mathematical analysis is beyond the scope of this book.

From the viewpoint of cellular engineering if we want to use this mathematical concept of cellular kinetics to study the cellular behaviour, we need to study the biological experiment that involves mRNA production and consequent protein synthesis. During this process, we collect the data at different points of time for a series of possible events (reaction kinetics). We can then use these data for optimising the parameters in the proposed mathematical model (Eqs. (10.10)). Once we use those parameters and perform the model analysis, we can then be able to demonstrate the system's behaviour. There are a number of optimisations techniques available to use in this context. This process is what is called as *model validation*.

10.3 Ontogenesis

Ontogenesis consists of three basic steps, e.g., *cell differentiation*, *cell proliferation* (increase the number of cell and the mass of the embryo), and *morphogenesis*. Cellular differentiation is governed by regulatory intra- and inter-cellular interactions that take place by the phenomena of molecular signalling, molecular recognition. The active movements of cells as a result of mechanochemical process, which is also simulated by molecular signalling, determine *morphogenesis*.

So far, we have developed an idea about the cellular kinetics that plays important part within a cell. An organ is a collection of many cells and extracellular matrix with a specific architectural structure and for a specific purpose. The rest of the chapter will deal with discussion of various cellular engineering techniques that may help readers to understand general concept of this complex mechanism. Differentiation at the early stage depends on two factors. The first one occurs inside the zygote due to non-uniform distribution of mass inside the cytoplasm. The second factor arises due to cell division.

The mathematical modelling of ontogenesis was first proposed by Turing [111]. The basic assumptions made by Turing are as follows:

(i) The position of a cell in a developing system is defined by its radius vector r_i.

(ii) The concentration gradients of the functional substances are significantly small compared to the size of the cell. Therefore, one can assume r_i is continuous.

(iii) State of the cell is defined by $x_j(r_i)$.

Hence, the equation for mass balance is given by

$$
\frac{\partial}{\partial \tau}
\begin{pmatrix}
x_1(r) \\
x_2(r) \\
\cdot \\
\cdot \\
\cdot \\
x_j(r) \\
\cdot \\
\cdot \\
\cdot \\
x_n(r)
\end{pmatrix}
=
\begin{pmatrix}
F_1\Big(x_1(r), x_2(r), \ldots x_j(r), \ldots x_n(r)\Big) \\
F_2\Big(x_1(r), x_2(r), \ldots x_j(r), \ldots x_n(r)\Big) \\
\cdot \\
\cdot \\
\cdot \\
F_j\Big(x_1(r), x_2(r), \ldots x_j(r), \ldots x_n(r)\Big) \\
\cdot \\
\cdot \\
\cdot \\
F_n\Big(x_1(r), x_2(r), \ldots x_j(r), \ldots x_n(r)\Big)
\end{pmatrix}
+
\begin{pmatrix}
D_{11} & 0 & 0 & \ldots & \ldots & 0 \\
0 & D_{22} & 0 & \ldots & \ldots & 0 \\
\cdot & \cdot & \cdot & \cdot & & \cdot \\
\cdot & \cdot & \cdot & \cdot & & \cdot \\
0 & 0 & 0 & \ldots D_{jj} & \ldots & 0 \\
\cdot & \cdot & \cdot & \cdot & & \cdot \\
0 & 0 & \ldots & \ldots & \ldots & 0 \\
0 & 0 & \ldots & \ldots & \ldots & D_{nn}
\end{pmatrix}
$$

$$(10.11)$$

where n is the number of various reactants.

It is also assumed that the different macroscopic structure arise due to different stable steady states. Martinez [66] proposed a different model of linear array of cells by modifying Turing model. According to Martinez, a cell divides when the concentrations of some substances within the cell reached their threshold values. The rate change of these concentrations depend upon two substances within the cell. Martinez called it *morphogen*. One of the morphogens can diffuse between the cells and finally form an inhomogeneous distribution of matter among the cells.

Most of the studies of morphogenesis and pattern formation have been based on the models developed by Gierer and Meinhardt [31] and others. All of these models have followed the Turing concept. Further analysis of these models have been made by Nicolis and Auchmuty [78], Saha and Tapaswi [93], Hunding and Saha [42], Auchmuty and Nicolis [4] and others. We will now discuss a mathematically designed example of pattern formation and morphogenesis as proposed by Saha [91].

Consider an example of mathematical modelling framework during embryogenesis involving mRNA regulator, which is itself a protein, perhaps an enzyme, and morphogen taking into account the self and cross-diffusion of morphogen. By self-diffusion (passive transport), we mean that the diffusive substance moves from higher concentration to lower concentration while in case of cross-diffusion (counter transport), the diffusive substance moves towards the concentration gradient of another substance. The term negative cross-diffusion implies that the diffusive substance moves towards the higher concentration of another substance which is sometimes called active counter transport. A chemical basis of this line of thought has been reported by Jorne [48], Marchbanks [63], and others.

The homogeneous mathematical model, considered here, involves mRNA(X), regulator(Y), which is itself a protein, perhaps an enzyme, and morphogen (Z) and can be expressed as

$$\frac{dX}{dT} = \frac{1}{1+Y^\rho} - \gamma_1 X, \tag{10.12a}$$

$$\frac{dY}{dT} = X - \gamma_2 Y, \tag{10.12b}$$

$$\frac{dZ}{dT} = Y - \gamma_3 Z. \tag{10.12c}$$

Incorporating the self and cross diffusions of the morphogen, the model equations can further be modified as

$$\frac{d\phi}{dt} = F(\phi) + \gamma_3 H(\phi) + D\nabla^2 \phi \quad \text{with zero flux boundary condition} \tag{10.13}$$

where $\nabla^2 \equiv$ Laplace operator in cylindrical coordinates (note: geometry depends on the shape of the embryo). In this example the shape of the embryo considered to be cylindrical in form such that

$$\nabla^2 \equiv \frac{\partial^2}{\partial r^2} + \left(\frac{1}{r}\right)\frac{\partial}{\partial r} + \left(\frac{1}{r^2}\right)\frac{\partial^2}{\partial \theta^2} + \frac{\partial^2}{\partial z^2}, \tag{10.14}$$

where $r \geq 0, 0 \leq \theta \leq 2\pi$ and $-\infty < z < +\infty$. Here, ρ is the Hill's coefficient arising from cooperative enzyme kinetics, and $\gamma_i (i = 1, 2, 3)$ are system parameters. We have denoted here

$$\phi = \begin{pmatrix} X \\ Y \\ Z \end{pmatrix}, F(\phi) = \begin{pmatrix} \frac{1}{1+Y^\rho} - \gamma_1 X \\ X - \gamma_2 Y \\ Y \end{pmatrix}, H(\phi) = \begin{pmatrix} 0 \\ 0 \\ -Z \end{pmatrix} \text{ and } D = \begin{pmatrix} 0 & 0 & 0 \\ 0 & 0 & D_{23} \\ 0 & 0 & D_{33} \end{pmatrix}.$$

$$\tag{10.15}$$

We now consider a cylindrically shaped embryo as a model, because Drosophila egg, a system that has been thoroughly studied, is of nearly cylindrical shape.

10.3.1 *Steady State and Linearisation*

The steady state $w_0 = (X_0, Y_0, Z_0)^t$ of the system (10.13) can be found from the following set of equations

$$\begin{pmatrix} \frac{1}{1+Y^\rho} - \gamma_1 X \\ X - \gamma_2 Y \\ Y - \gamma_3 Z \end{pmatrix} = \begin{pmatrix} 0 \\ 0 \\ 0 \end{pmatrix}. \tag{10.16}$$

The linearised version of the system (10.13) around the steady state w_0 can be obtained as

$$\frac{dU}{dt} = \partial F(w_0)U + \gamma_3 \partial H(w_0)U + D\nabla^2 U, \tag{10.17}$$

where

$$U = \begin{pmatrix} X - X_0 \\ Y - Y_0 \\ Z - Z_0 \end{pmatrix}. \tag{10.18}$$

In nature, there are many evidence of almost cylindrically shaped embryos with one more pre-assigned gradient of certain known gene products (Johnston and Volhard [49]). In this example, we shall not outline in detail about the mechanism for the emergence of these gradients but recent experiments on Drosophila suggest that they arise from the maternal genes.

For simplicity, consider there exists only one pre-assigned gradient along the anterior-posterior axis of the embryo. In reality there may exist more gradients along the anterior-posterior axis. As an example in Drosophila, a gradient from a gene bicoid is set up from the anterior pole as well as a gradient in the Oscar group of genes is set up from the posterior pole. In this context, it is assumed that this pre-assigned gradient exerts an effect only on the diffusion term of morphogen (Z), although it can affect the reaction kinetics. Incorporating all these facts, the system (10.13) becomes

$$\frac{d\phi}{dt} = F(\phi) + \gamma_3 H(\phi) + (D + \eta \Theta D')\nabla^2 \phi \tag{10.19}$$

with zero flux boundary condition where

$$D' = \begin{pmatrix} 0 & 0 & 0 \\ 0 & 0 & 0 \\ 0 & 0 & D_{33} \end{pmatrix}, \tag{10.20}$$

η is a real parameter and Θ is the pre-assigned gradient. As the steady state of the system (10.19) is the same as steady state of the system (10.13), therefore the linearised version of (10.19) around the same steady state w_0 can be expressed as

$$\frac{dU}{dt} = \partial F(w_0)U + \gamma_3 \partial H(w_0)U + (D + \eta \Theta D')\nabla^2 U, \tag{10.21}$$

where

$$\partial F(w_0) = \begin{pmatrix} -\gamma_1 & k & 0 \\ 1 & -\gamma_2 & 0 \\ 0 & 1 & 0 \end{pmatrix}, \partial H(w_0) = \begin{pmatrix} 0 & 0 & 0 \\ 0 & 0 & 0 \\ 0 & 0 & -1 \end{pmatrix}. \tag{10.22}$$

Inhomogeneous concentration pattern emerges as a solution of the stationary case of the system in (10.17),

$$\partial F(w_0)U + \gamma_3 \partial H(w_0)U + D\nabla^2 U = 0 \quad \text{with zero flux boundary condition.} \tag{10.23}$$

Let us denote $M = \partial F(w_0) + \gamma_3 \partial H(w_0)$, then Eq. (10.23) becomes

$$MU + D\nabla^2 U = 0 \qquad (10.24)$$

where

$$M = \begin{pmatrix} -\gamma_1 & k & 0 \\ 0 & -\gamma_2 & 0 \\ 0 & 1 & -\gamma_3 \end{pmatrix} \quad \text{and} \quad k = \rho\gamma_1^2\gamma_2^2 Y_0^{\rho+1}. \qquad (10.25)$$

System (10.23) has a non-trivial solution $U(r) \neq 0$, if the parameter attains a critical value say, γ_3^c. This happens when

$$\det|M_c - \kappa_{nkj}^2 D| = 0, \qquad (10.26)$$

where M_c is obtained from M, replacing γ_3 by γ_3^c. The second term in Eq. (10.26) comes from Helmholtz equation $\nabla^2 \psi_{nkj} = -\kappa_{nkj}^2 \psi_{nkj}$ with zero-flux boundary condition $\frac{\partial \psi_{nkj}}{\partial r} = 0$, for $r = R$, radius of the cylindrically shaped embryo, ψ_{nkj} and κ_{nkj}^2 are eigenvector and eigenvalues of the Laplacian. From Eq. (10.26) we get

$$\gamma_3^c = \kappa_{nkj}^2 D_{33} \left\{ \frac{\sigma}{\gamma_2(1 + \gamma_1\gamma_2\rho Y_0^{\rho+1})} - 1 \right\} \text{ where } \sigma = \frac{D_{23}'}{D_{33}} > 0 \text{ and } -D_{23} = D_{23}' > 0.$$
$$(10.27)$$

10.3.2 *Pattern in Cylindrical Shaped Model Embryo*

For each set of (n, k, j) values, one can find a unique value of $\gamma_3 = \gamma_3^c$ for which the system (10.23) has non-trivial solution

$$U \equiv U_L = \delta\psi_{nkj} \qquad (10.28)$$

where

$$\psi_{nkj} = \cos\frac{k\pi}{h} z J_n(x_{nj}\frac{r}{R}) \left\{ \begin{array}{c} \cos n\phi \\ \sin n\phi \end{array} \right\}, \qquad (10.29)$$

and $n, k = 0, 1, \ldots$, and $j = 1, 2, \ldots$. Here, $J_n(x)$ represents the Bessel's function of order n, x_{nj} is the jth root of $J_n'(x) = 0$, and $\kappa_{njk}^2 = \left(\frac{k\pi}{h}\right)^2 + \left(\frac{x_{nj}}{R}\right)^2$, h and R are the height and radius of the cylindrically shaped embryo. The solution to the system (10.23) is given by

$$U_L = \nu_{nkj} \begin{pmatrix} \delta_{nkj}^1 \\ \delta_{nkj}^2 \\ \delta_{nkj}^3 \end{pmatrix} \cos\left(\frac{k\pi}{h}z\right) J_n\left(x_{nj}\frac{r}{R}\right)\cos n\phi \qquad (10.30)$$

where $\delta^t = \left(\delta_{nkj}^1 \delta_{nkj}^2 \delta_{nkj}^3\right)^t$ can be obtained from $\left[M_c - \kappa_{nkj}^2 D\right]\delta = 0$ and ν_{nkj}, is the scaling factor which can be obtained from the inner product $\langle U_L, U_L \rangle = 1$.

Fig. 10.3: The first pattern which occurs at the gastrula stage during embryogenesis.

A real pattern actually occurs in gastrula stage of an embryogenesis of a large number of species, and can be obtained for a specific values of $(n, k, j) = (0, 0, 3)$. The earliest occurring patterns, i.e., *endoderm, mesoderm* and *ectoderm* in the cylindrically shaped embryo at about gastrula stage of the embryonic development correspond to the solution of (10.30) when $n = 0$, $k = 0$, $j = 3$, and for $\gamma_1 = 1, \gamma_2 = 0.5, \rho = 2, \sigma = 2, h = 6$ and $R = 1$ (height and radius of the cylindrically shaped embryo). The y-axis through the wavy curve is the base line representing the steady-state value of the regulator (Y). The central large deflection of Y above Y0 (Y $\gg$ Y0) in the xy-plane switches on the genes responsible for the formation of endodermal tissues, the downward deflection (Y < Y0) on both sides of the former switches on the genes corresponding to the mesodermal tissues. The small deflection above the base line (Y > Y0) on the extreme sides switches on the genes corresponding to the ectodermal tissues (see Saha [91]).

These three layers can be related to biologically obtained structure of an embryogenesis. For example, *ectoderm* is the most outer layer and biologically, this layer differentiates to form the nervous system, tooth enamel, the epidermis, lining of mouth, anus, nostrils, sweat glands, hair and nails. The inner most layer biologically concurs with *mesoderm* that forms *mesenchyme* (connective tissue), *mesothelium*, non-epithelial blood cells and *coelomocytes*. The inner most layer can be related to *endoderm*. Cells in this layer form the epithelial lining of multiple systems.

To illustrate this complex mechanism, consider an example of an egg of *Tribolium*, flour beetle (Fig. 10.4). The newly laid egg has only one nucleus and the rest are fat and proteins. This single DNA multiply many times and form a single cell layer on the outside of the egg, known as *blastoderm* stage (http://www.science-explained.com/theory/extraembryonic-development/). Afterwards, this single layer of cells in the periphery of the egg becomes three different types of cells (Fig. 10.5).

During gastrulation the cells of the blastula stage are dramatically rearranged and

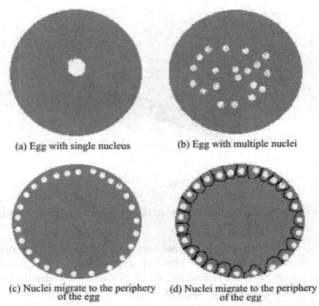

(a) Egg with single nucleus   (b) Egg with multiple nuclei

(c) Nuclei migrate to the periphery (d) Nuclei migrate to the periphery
of the egg   of the egg

Fig. 10.4: Egg with single and multiple nuclei.

Fig. 10.5: Three different types of cells.

relocated to new positions. The following figures (Fig. 10.6) of flour beetle embryo may give an idea of subsequent stages of embryonic development. As we can see the gastrulation is a well-organised sequential process of embryonic development guided by a set of coordinated complex mechanisms.

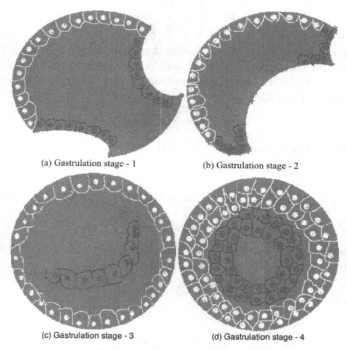

(a) Gastrulation stage - 1   (b) Gastrulation stage - 2

(c) Gastrulation stage - 3   (d) Gastrulation stage - 4

Fig. 10.6: Gastrulation stages.

It was Morgan [74] who first introduced the concept of morphogen in his pioneering work on Drosophila embryo. He first proposed that gradients of determinative substances inspire specific regeneration. Subsequently, Wolpert [119] advanced this morphogen concept using his famous French flag model. He introduced the concept of diffusive morphogenetic gradient that produces a long range pattern by subdividing a homogeneous mass into domain of different targeted gene expression.

Volhard [49] was the first person who identified a morphogen gradient experimentally, known as, *Bicoid*, one of the transcription factors present in a gradient in the Drosophila syncytial embryo. We have seen the existence of preassigned morphogenetic gradient in Drosophila embryo. We will now see analytically how the preassigned gradient influence the growth dynamics and/or pattern during embryogenesis.

The system (10.19) can be rewritten as

$$\frac{d\phi}{dt} = F(\phi) + \gamma_3 H(\phi) + (D + \eta\Theta D')\nabla^2\phi \qquad \text{with zero flux boundary condition}$$
(10.31)

where

$$D' = \begin{pmatrix} 0 & 0 & 0 \\ 0 & 0 & 0 \\ 0 & 0 & D_{33} \end{pmatrix},$$

η is a real parameter and Θ is the pre-assigned gradient. For $\eta = 0$, the above system reduces to the original system (10.13) without any preassigned gradient. Accordingly, for small $\eta \neq 0$, i.e., for the preassigned gradient Θ, the system parameter γ_3 has been perturbed, and can be expressed as

$$\gamma_3 = \gamma_3^c + \omega$$
(10.32)

and the corresponding linear solution takes the form

$$U_L = U_L^{(0)} + \eta U_L^{(1)} + \eta^2 U_L^{(2)} + \dots.$$
(10.33)

It is assumed that for small $\eta \neq 0, \omega$ can be expressed as an asymptotic expansion of η, i.e.,

$$\omega = \eta\omega^{(1)} + \eta^2\omega^{(2)} + \eta^3\omega^{(3)} + \dots.$$
(10.34)

Inserting Eqs. (10.32)-(10.34) in the linearised system (10.23) and then equating the like powers of η, a system of equations will generate

$$\left[M^{(0)} + D\nabla^2\right] U_L^{(0)} = 0,$$
(10.35)

$$\left[M^{(0)} + D\nabla^2\right] U_L^{(1)} = -\left[M^{(1)} + \Theta D'\nabla^2\right] U_L^{(0)},$$
(10.36)

$$\left[M^{(0)} + D\nabla^2\right] U_L^{(2)} = -\left[M^{(1)} + \Theta D'\nabla^2\right] U_L^{(1)} - M^{(2)}U_L^{(0)},$$
(10.37)

and so on. Functions $M^{(i)}$ $(i = 0, 1, 2, \dots)$ are defined as follows

$$M^{(0)} = \partial F(w_0) + \gamma_3^c \partial H(w_0),$$
(10.38)

and

$$M^{(i)} = \begin{bmatrix} 0 & 0 & 0 \\ 0 & 0 & 0 \\ 0 & 0 & -1 \end{bmatrix} \omega^{(i)} \partial H(w_0).$$
(10.39)

The adjoint problem of (10.35) can be written as

$$\left[M^{(0)} + D\nabla^2\right]^* U_L^* = 0$$
(10.40)

such that

$$\left\langle U_L^*, [M^{(0)} + D\nabla^2]U_L^{(0)} \right\rangle = \left\langle [M^{(0)} + D\nabla^2]U_L^*, U_L^{(0)} \right\rangle.$$
(10.41)

The solution to the above adjoint problem is given by

$$U_L^* = \delta^* \cos\frac{k\pi}{h} z J_n(x_{nj}\frac{r}{R}) \cos n\phi \tag{10.42}$$

where

$$\delta^* = \begin{bmatrix} 1 \\ \gamma_1 \\ \frac{\gamma_1 K_{nkj}^2 D_{23}'}{\gamma_3^c + K_{nkj}^2 D_{33}} \end{bmatrix}. \tag{10.43}$$

Instead of biochemical substance we assume a periodic signal, as preassigned gradient originates from a reference point. Here, the reference point is the anterior-pole of the cylindrically shaped embryo. The periodic signal is propagating from anterior-pole towards posterior-pole of the embryo. A different type of preassigned gradient along different axis of the embryo can also be tried. In this example, we assume

$$\Theta = \cos\frac{\alpha\pi}{h} z, \qquad \alpha = 1, 2, 3\ldots. \tag{10.44}$$

The embryo without preassigned gradient has the critical bifurcating parameter at which pattern evolve is $\gamma_3 = \gamma_3^c$. Due to the presence of preassigned gradient, this critical parameter is now being shifted to a new position $\gamma_3 = \gamma_3^c + \eta^2\omega^{(2)}$, where

$$\omega^{(2)} = -\frac{\left[\delta_{n(k-\alpha)j}^{(13)} K_{n(k-\alpha)j}^2 + \delta_{n(k+\alpha)j}^{(13)} K_{n(k+\alpha)j}^2\right] D_{33}}{2\nu_{nkj}^{(1)} \delta_{nkj}^{(3)}}. \tag{10.45}$$

The corresponding linear solution is given by

$$U_L = J_n(x_{nj}\frac{r}{R}) \cos n\phi \left\{ \nu_{nkj}^{(1)} \begin{bmatrix} \delta_{nkj}^{(1)} \\ \delta_{nkj}^{(2)} \\ \delta_{nkj}^{(3)} \end{bmatrix} \cos\frac{k\pi}{h} z + \eta \begin{bmatrix} \delta_{n(k-\alpha)j}^{(11)} \\ \delta_{n(k-\alpha)j}^{(12)} \\ \delta_{n(k-\alpha)j}^{(13)} \end{bmatrix} \cos\frac{(k-\alpha)\pi}{h} z \right.$$

$$\left. + \eta \begin{bmatrix} \delta_{n(k+\alpha)j}^{(11)} \\ \delta_{n(k+\alpha)j}^{(12)} \\ \delta_{n(k+\alpha)j}^{(13)} \end{bmatrix} \cos\frac{(k+\alpha)\pi}{h} z \right\}. \tag{10.46}$$

We thus see that embryo without preassigned gradient may develop a realistic pattern for certain combination of values of $(n = 0, k = 0, j = 3)$ which is considered to be the gastrula stage of embryonic development. The essential idea of considering preassigned gradient within embryo is to explore how one stage of pattern can be altered when a preassigned gradient exerts some effects on it. We shall now investigate the effect of preassigned gradient Θ on a realistic pattern at gastrula stage of the embryo.

Case 1: when $\alpha = 1$, $\Theta = \cos\frac{\pi}{h} z$, the corresponding pattern will become as shown in Fig. 10.7 (see Saha [92]). Due to the effect of this preassigned gradient, the

Fig. 10.7: For $\alpha = 1$, the preassigned gradient becomes $\Theta = \cos \frac{\pi}{6} z$.

triple structure - *ectoderm*, *mesoderm* and *endoderm* tends to vanish at or near the posterior end of the cylindrical embryo where $\gamma_1 = 1, \gamma_2 = 0.5, \rho = 2, \sigma = 2, h = 6$ and $R = 1$.

Case 2: when $\alpha = 2$, $\Theta = \cos \frac{2\pi}{h} z$, the corresponding pattern will become as shown in Fig. 10.8. For $\alpha = 2$, the preassigned gradient becomes $\Theta = \cos \frac{\pi}{3} z$. Due

Fig. 10.8: (a) Outer view of the embryo, (b) Cross-sectional view of the embryo.

to the effect of this preassigned gradient, the triple structure - *ectoderm*, *mesoderm* and *endoderm* vanishes just at the middle of the embryo where $\gamma_1 = 1, \gamma_2 = 0.5, \rho = 2, \sigma = 2, h = 6$ and $R = 1$.

Case 3: when $\alpha = 12$, $\Theta = \cos \frac{12\pi}{h} z$, the corresponding pattern will become as shown in Fig. 10.9. For $\alpha = 12$, the preassigned gradient becomes $\Theta = \cos 2\pi z$. Due to the effect of this preassigned gradient, the triple structure - *ectoderm*, *mesoderm* and

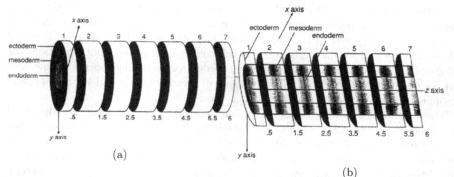

Fig. 10.9: (a) Outer view of the embryo, (b) Cross-sectional view of the embryo.

endoderm repeated seven times yielding the 7-stripe structure within the cylindrical embryo where $\gamma_1 = 1, \gamma_2 = 0.5, \rho = 2, \sigma = 2, h = 6$ and $R = 1$.

10.4 Concluding Remarks on Tissue Engineering (TE) and Cellular Engineering (CE)

It is well known that tissue engineering is a new interdisciplinary field dealing with the principles of engineering and life sciences towards the development of biological substitutes that restore, maintain and improve tissue function, whereas, cellular and molecular bioengineering involves the mechanical, electrical and chemical processes of the human cell and tries to explain how cellular behaviour arises from molecular-level interactions.

In the last two chapters our focus is to understand cell biological process and physiological aspects of developmental biology. For this, we need to study (i) range of different specifically designed bio-reactors, (ii) the way they are cultured (not pre-cultured and just seeded and immediately implanted, or cultured in static conditions or in an all range of different specifically designed bio-reactors), (iii) to the design of adequate scaffolds, including the selection of adequate materials and routes to process them, the respective porosity, interconnectivity, surface characteristics, hydrophilicity, etc., and (iv) the use of adequate animal models that will allow us to test the efficacy of different tissue engineering approaches and the potential of different constructs (distinct combinations of scaffolds/cells/in vitro culturing conditions). Finally, all the related ethical considerations, especially when considering the use of stem cells (and potentially embryonic stem cells) and animal studies, as well as the need for proper standards must be taken into account. In depth knowledge of cellular engineering is an important part of this TE process in order to find the appropriate cell source and adopting a right environment for cellular growth kinetics.

Tissue engineering (TE) as well as cellular engineering (CE) are relatively young fields that aim at repairing, regenerating, or replacing damaged tissues with cellularised constructs grown in the laboratory. The arsenal of TE comprises those of cell biology, needed for finding a suitable cell source and for assuring conditions for cell growth, those of biomaterial chemistry, needed for preparing biocompatible support for anchorage dependent cells, and those of physiology, needed for maintaining a biomimetic, organ-specific environment in vitro.

The promise of solving the problem of transplantable organ shortage is only one of the engines that drive TE research. Several laboratories and biotechnology companies have already developed organ modules that are used for tissue repair and regeneration. Moreover, functional subunits of human organs are viewed as a viable alternative for animal testing of new drugs. As cells divide, differentiate, and organise into tissues and organs during embryonic development, they produce a tissue-specific mixture of interconnected protein filaments, the extracellular matrix (ECM). Besides offering a three-dimensional (3D) support for cells, the ECM is also involved in cell signalling. Cell-ECM interactions play a crucial role in the function and structural integrity of the tissue. Instead of attempting to reproduce the very complex composition of ECM in vitro, TE and CE aim at fabricating supportive scaffolds and identifying culture conditions that promote ECM production by cells. The basic tool in this endeavour is a bioreactor that allows for the controlled conditioning of the engineered construct. As the field of TE and CE evolved, comparative studies between bioreactors have pointed to temperature control, gas exchange, mass transfer, shear stress, and other mechanical stimuli as essential factors of development and maturation.

GLOSSARY

AIDS — acquired immunodeficiency syndrome.

Aorta — the main trunck from which the systemic arteries originate.

Atrium — an upper chamber on either side of the heart.

Auscultation — listening for sounds produced within the body.

Axon — the long outgrowth of the body of a nerve cell.

Cardiac output (CO) — the steady rate at which blood is ejected by the heart (5 litres min^{-1} at rest for man).

Catheter — a tubular instrument passed through body channels for withdrawal or introduction of body fluids.

Cineangiography — the photographic recording of fluoroscopic images of the organ by motion picture techniques.

Developmental biology — study of cell growth, differentiation and morphogenesis.

Diabetes Mellitus — the unusual viscous properties of blood.

Diastole — the phase of the cardiac cycle in which the heart relaxes between contractions.

DNA — deoxyribonucleic acid found in chromosomes.

Echocardiography — recordings of internal structures of the heart by the echo obtained from ultrasound beams.

ECM — extracellular matrix.

Electrocardiography — the graphic recording of the electric currents generated by the heart. Tracing is called ECG.

Endothelium — a delicate membrane lining the heart.

Epigenesis — fundamental problem of embryology.

Erythrocyte — a red blood cell, or corpuscle.

ESR — erythrocyte sedimentation rate.

Flowmeter — an apparatus for measuring the rate of flow of liquids or gases.

GAG — glycosaminoglycans.

Gastrulation — process of embryonic development.

HIV — human immunodeficiency virus.

Hypertension — persistently high blood pressure.

Inviscid fluid — non-viscous or viscosity neglected.

In vitro — within a glass.

In vivo — within the living body.

$\kappa - \epsilon$ model — model used for the solution of turbulent flow problems.

Laminar flow — fluid which flows in laminae (purely viscous flow).

Lumen — the cavity or channel within a tubular organ, as a blood vessel or the intestine.

Metabolism — the process by which in any living body, human or plants, food is converted into energy.

Monomers — basic, non-repeating units used in cell-biology.

Morphogenesis — spatial distribution of cells giving rise to tissue and organs.

Myocardium — the thickest layer of the heart wall.

Myocardial infarction — heart attack.

Noninvasive — without any surgical intervention.

PEG — polyethylene glycol

PG — proteoglycan

Plasma — an aqueous solution of electrolytes and organic substances.

RBC — red blood cells or erythrocytes.

RNA — ribonucleic acid found in chromosomes.

Stenosis — narrowing of a blood passage or opening.

Systole — the period of the cardiac cycle in which the heart contracts.

Thrombosis — formation or presence of blood clots inside a blood vessel.

Ultrasound — mechanical energy of a frequency greater than 20,000 cycles per second.

Vasoconstrictor — causing constriction of the blood vessels.

Vasodilator — causing dilation of blood vessels.

Ventricle — a small cavity or chamber, as in the brain or heart.

WBC — white blood cells or leucocytes.

BIBLIOGRAPHY

The list of references included here is meant to be a helpful guide through the literature rather than a complete catalogue. My sincere apologies to the many authors whose work is not in the list. The references are listed alphabetically according to the first author. There are several related references, which are ordered according to date of publication.

1. P. H. Abraham, B. M. Logan, R. T. Hutchings and J. D. Spratt, *McMinn's The Human Skeleton*, 2nd edition (Mosby, Cambridge, UK, 2006).

2. A.D. Au and H.S. Greenfield, *Comput. Biol. Med.* **4** (1975) p. 279.

3. A.D. Au and H.S. Greenfield, *Comput. Biomed. Res.* **10** (1977) p. 165.

4. J.F.G. Auchmuty and G. Nicolis, *Bull. Math. Bio.* **38** (1976) p. 325.

5. C.P. Bailey, *Dis. Chest* **15** (1949) p. 377.

6. D.W. Baker and R.E. Daigle, in *Cardiovascular Flow Dynamics and Measurements*, ed. N.H.C. Hwang and N.A. Normann (University Park Press, Baltimore, 1977).

7. A.F.M. Barton, *Handbook of Polymer-Liquid Interaction Parameters and Solubility Parameters*, (CRC Press, Boca Raton, 1990).

8. B.J. Bellhouse and L. Talbot, *J. Fluid Mech.* **35** (1969) p. 721.

9. D.H. Bergel, *Cardiovascular Fluid Dynamics, Vol. 2* (Academic Press, London and New York, 1972) p. 158.

10. V.O. Björk, *Scand. J. Thorac. Cardiovasc. Surg.* **3** (1969) p. 1.

11. P.L. Blackshear, in *Chemistry of Biosurfaces, Vol. 2*, ed. M.L. Hair (Marcel Dekker, New York, 1972) p. 523.

12. J.J. Blum, *Am. J. Physiol.* **198** (1960) p. 991.

13. J.D. Bronzino, *Biomedical Engg. Handbook*, 3rd edition (Taylor and Francis, USA, 2006).

14. E. Buckingham, *Phys. Rev.* **4** (1914) p. 345.

15. J. Carr, *Appl. Math. Scien. Vol. 35* (Springer-Verlag, New York, 1981).

16. M. Casson, in *Rheology of Dispersive Systems*, ed. C.C. Mills (Pergamon, Oxford, 1959) p. 84.

17. S.E. Charm and G.S. Kurland, *Nature* **206** (1965) p. 617.

18. C.W. Davies, *J. Comput. Phys.* **39** (1981) p. 164.

19. T. Doenst, *Ann. Thorac. Surg.* **87** (2009) p. 1187.

20. M. Doi, *Introduction to Polymer Dynamics* (Oxford Press, 1996).

21. P.T. English and C. Moore, *MRI for Radiographers* (Springer-Verlag, UK, 1995).

22. M.S. Engelman, S.E. Moskowitz and J.B. Borman, *J. Thorac. Cardiovasc. Surg.* **79** (1980) p. 402.

23. R.S. Figliola and T.J. Muller, *A.S.M.E. J. Biomech. Eng.* **99** (1977) p. 173.

24. P. J. Flory, *Principles of Polymer Chemistry* (Cornell University Press, Ithaca, NY, 1953).

25. J.H. Forrester and D.F. Young, *J. Biomech.* **3** (1970) p. 297.

26. M. Friedman, *J. Eng. Math.* **6** (1972) p. 285.

27. D.L. Fry, *Circ. Res.* **22** (1968) p. 165.

28. Y.C. Fung, *Biomechanics: Mechanical Properties of Living Tissues* (Springer-Verlag, New York, 1981).

29. Y.C. Fung, *Biodynamics Circulation* (Springer-Verlag, New York, 1984).

30. Y.C. Fung, *Biomechanics: Motion, Flow, Stress and Growth* (Springer-Verlag, New York, 1990).

31. A. Gierer and H. Meinhardt, *J. Cell. Sci.* **15** (1974) p. 321.

32. B.C. Goodwin, *Temporal Organization in Cells* (Academic Press, New York, 1973).

33. H.C. Groen, *J. Biomech.* **43** (2010) p. 2332.

34. D.N. Ghista, *Proc. 2nd Int. Conf. Mech. Med. Biol.* (Osaka, Japan, 5-7 June, 1980).

35. G. Hagen, *Poggendorffs Ann.* **46** (1839) p. 423.

36. Y.A.M. Haggag, *J. Engg. Med.* **204** (1990) p. 111.

37. D.E. Harken, H.S. Soroff, W.J. Taylor, A.A. Lefemine, S.K. Gupta and S. Lunzer, *J. Thorac. Cardiovasc. Surg.* **40** (1960) p. 744.

38. J.D. Hellums and C.H. Brown, in *Cardiovascular Fluid Dynamics*, ed. N.H.C. Hwang and N.A. Normann (University Park Press, Baltimore, 1977) p. 799.

39. W.H. Herkes and J.R. Lloyd, *A.S.M.E. J. Biomech. Eng.* **103** (1981) p. 267.

40. C.A. Hufnagel, W.P. Harvey, P.J. Rabil and T.F. McDermott, *Surgery* **35** (1954) p. 673.

41. C.A. Hufnagel and P.W. Conrad, *Surgery* **57** (1965) p. 205.

42. A. Hunding and A.K. Saha, Pattern generation in reaction-diffusion systems (Turing structures), in *Models of Self-Organization on Complex Systems*, Vol. 64, 142–147, Mathematical Research, 1991.

43. T.C. Hung, R.M. Hochmuch, J.H. Joist and S.P. Sutera, *Trans. Am. Soc. Artif. Intern. Organs* **22** (1976) p. 285.

44. S.R. Idelsohn, L.E. Costa and R. Ponso, *J. Biomech.* **18** (1985) p. 97.

45. F. Jacob and J. Monod, *Cold Spring Harber Symp. Quant. Biol. Cell. Regul. Mech.*, **26** (1961) p. 193.

46. G.B. Jeffery, *Proc. Roy. Soc. A* **102** (1922) p. 161.

47. W.P. Jones and B.E. Launder, *Int. J. Heat Mass Transfer* **16** (1973) p. 119.

48. J. Jorne, *J. Theor. Biol.* **55** (1975) p. 529.

49. D.S. Johnston and N.C. Volhard, *Cell* **68** (1992) p. 201.

50. J.N. Kapur, *Mathematical Models in Biology and Medicine* (Affiliated East-West Press PVT. Ltd., India, 1985).

51. R.L. Kaster and C.W. Lillehei, *Dig. 7th Int. Conf. Med. Biol. Eng.* (Stockholm, 1967) p. 387.

52. E.B. Kay, A. Suzuki, M. Demaney and H.A. Zimmerman, *Amer. J. Cardiol.* **18** (1966) p. 504.

53. J.P. Keener, S. Sircar and A.L. Fogelson, *SIAM J. Appl. Math.* **71**(3) (2011) p. 854.

54. J.P. Keener, S. Sircar and A.L. Fogelson, *Phys. Rev. E* **83**(4) (2011) p. 041802.

55. A. Krogh, *J. Phys.* **52** (1919) p. 457.

56. R. Langer and J.P. Vacanti, *Science* **260** (1993) p. 920.

57. J.F. LaDisa, *Am. J. Physiol. Heart Circ. Physiol.* **288**(5) (2005) p. H2465.

58. B.E. Launder and D.B. Spalding, *Comput. Methods Appl. Mech. Eng.* **3** (1974) p. 269.

59. L.B. Leverett, J.D. Hellums, C.P. Alfrey and E.C. Lynch, *Biophys. J.* **12** (1972) p. 257.

60. J. Lighthill, *Physiological Fluid Mechanics* (Springer-Verlag, New York, 1971).

61. J. Lighthill, *Mathematical Biofluid Dynamics* (SIAM, Philadelphia, U.S.A., 1975).

62. A.M. Malek, S.L. Alper and S. Izumo, *J. Amer. Med. Assoc.* **282**(21) (1999) 2035.

63. R.M. Marchbanks, *Membranes and ion Transport, Vol. 2* (Wiley, New York, 1970).

64. A. Maroudas, *Adult Articular Cartilage*, 2nd edition (Pitman Medical, Kent, England, 1979).

65. I. Marshall, *J. Biomech.* **37**(5) (2004) p. 679.

66. H.M. Martinez, *J. Theor. Biol.* **36** (1972) p. 479.

67. J. Mazumdar and K. Thalassoudis, *Med. Biol. Eng. Comput.* **21** (1983) p. 400.

68. J. Mazumdar, *An Introduction to Mathematical Physiology and Biology*, 2nd edition (Cambridge University Press, Cambridge, 1999).

69. J. Mazumdar, K.C. Ang and L.L. Soh, *J. Australas. Phys. Eng. Sci. Med.* **14** (1991) p. 65.

70. E.W. Merrill, A.M. Benis, E.R. Gilliland, T.K. Sherwood and E.W. Salzman, *J. Appl. Physiol.* **20** (1965) p. 954.

71. L. Michaelis and M.L. Menten, *Biochem. Z.* **49** (1913) p. 333.

72. C.D. Mobley and R.J. Stewart, *J. Comput. Phys.* **34** (1980) p. 124.

73. U. Morbiducci, *J. Biomech. Engg.* **132** (2010) p. 0910051.

74. T.H. Morgan, *Regeneration* (The Macmillan Company, New York, 1901).

75. B.E. Morgan and D.F. Young, *Bull. Math. Biol.* **36** (1974) p. 39.

76. R.M. Nerem, *Ann. Biomed. Engg.* **19** (1991) p. 529.

77. C.G. Nevaril, J.D. Hellums, C.P. Alfrey Jr. and E.C. Lynch, *AIChE J.* **15** (1969) p. 707.

78. G. Nicolis and J.F.G. Auchmuty, *Proc. Natn. Acad. Sci. U.S.A.* **71** (1974) p. 2748.

79. B.J. Noye, *Proc. Int. Conf. Numer. Simul. Fluid Dyn. Syst.* (Monash University, Melbourne, Australia, 1976) p. 1.

80. C.S. Peskin, *J. Comput. Phys.* **25** (1977) p. 220.

81. J. Poiseuille, *Compt. Rend.* **11** (1840) p. 961.

82. J. Poiseuille, *Compt. Rend.* **12** (1841) p. 112.

83. A.H. Reddi, *Nature Biotech.* **16** (1998) p. 247.

84. N. Reshevsky, *Mathematical Biophysics*, 3rd edition (Dover, New York, 1961).

85. H. Reul, M. Giersiepen and E. Knott, in *Heart Valve Engineering* (MEP, London, U.K., 1986).

86. M.R. Roach and A.C. Burton, *Can. J. Biochem. Physiol.* **35** (1957) p. 681.

87. P.J. Roache, *Computational Fluid Dynamics* (Hermosa, Albuquerque, New Mexico, 1976).

88. W. Rodi, *Turbulence Models and their Application to Hydraulics — A State of the Art Review* (International Association for Hydraulic Research, Delft, 1980).

89. R. Rosen, *Foundation of Mathematical Biology Vol. I, II & III* (Academic Press, New York, 1972).

90. S.I. Rubinow and J.B. Keller, *J. Theor. Biol.* **35** (1972) p. 299.

91. A.K. Saha, *Math. Comp. Model.* **24** (1996) p. 49.

92. A.K. Saha, *A new study on some bio-engineered tissues by knowledge based tissue engineering*, PhD thesis (2007).

93. A.K. Saha and P.K. Tapaswi, *App. Math. Model.* **13** (1989) p. 374.

94. A.K. Saha, J. Mazumdar and S.S. Kohles, *Ann. Biomed. Engg.* **32** (2004) p. 871.

95. A.K. Saha and J. Mazumdar, *IEEE Trans. Nanobiosci.* **2** (2003) p. 89.

96. T. Schenkel, *Ann. Biomed. Engg.* **37**(3) (2009) p. 503.

97. H. Schlichting, *Boundary Layer Theory*, 6th edition (McGraw-Hill, New York, 1968).

98. E. Schrödinger, *What is Life?* (Macmillan, New York, 1945).

99. G.W. Scott Blair, *Nature* **184** (1959) p. 354.

100. S. Sircar, J.P. Keener and A.L. Fogelson, *J. Chem. Phys.* **138** (2013) p. 014901.

101. H.S. Souttar, *Brit. Med. J.* **69** (1925) p. 603.

102. A. Starr and M.L. Edwards, *Ann. Thorac. Surg.* **154** (1961) p. 726.

103. P.D. Stein and H.N. Sabbah, *Circ. Res.* **39** (1976) p. 58.

104. D.M. Stevenson and A.P. Yoganathan, *J. Biomech.* **18** (1985) p. 899.

105. D.M. Stevenson and A.P. Yoganathan, *J. Biomech.* **18** (1985) p. 909.

106. S.P. Sutera and M.H. Mehrjardi, *Biophys. J.* **15** (1975) p. 1.

107. P.K. Tapaswi and A.K. Saha, *Bull. Math. Biol.* **48** (1986) p. 213.

108. K. Thalassoudis, *Numerical studies of flow through prosthetic heart valves*, Doctoral Thesis, University of Adelaide, Australia, 1987.

109. K. Thalassoudis, J. Mazumdar, B.J. Noye and I.H. Craig, *Med. Biol. Eng. Comput.* **25** (1987) p. 173.

110. J.F. Thompson and Z.U.A. Warsi, *J. Comput. Phys.* **47** (1982) p. 1.

111. A.M. Turing, *Phil. Trans. R. Soc.* **237** (1952) p. 37.

112. F.N. Underwood and T.J. Mueller, *A.S.M.E. J. Biomech. Eng.* **99** (1977) p. 91.

113. F.N. Underwood and T.J. Mueller, *A.S.M.E. J. Biomech. Eng.* **101** (1979) p. 198.

114. I. Villanueva, S.K. Gladem, J. Kessler and S.J. Bryant, *Mat. Biol.* **29** (2010) p. 51.

115. M.V. Volkenshtein, *Biophysics* (MIR Publishers, Moscow, 1983).

116. J.G. Webstar, *Encyclopedia of Medical Devices and Instrumentation, Vol. 3* (Wiley Interscience, New York, 1988).

117. S.L. Weinberg, in *The Ureter*, ed. Bergman (Harper and Row, U.S.A., 1967) p. 48.

118. C.G. Wilson, L.J. Banossar and S.S. Kohles, *Arch. Biochem. Biophys.* **408**, (2002) p. 246.

119. L. Wolpert, *J. Theor. Biol.* **25** (1969) p. 1.

120. K.K.L. Wong, R.M. Keslo, S.G. Worthley, P. Sanders, J. Mazumdar and D. Abbott, *Ann. Biomed. Engg.* **37**(8) (2009) p. 1495.

121. K.K.L. Wong, R.M. Keslo, S.G. Worthley, P. Sanders, J. Mazumdar and D. Abbott, *PLoS ONE* **4**(5) (2009) p. e5688.

122. K.K.L. Wong, R.M. Keslo, S.G. Worthley, P. Sanders, J. Mazumdar and D. Abbott, *Int. J. Art. Organs* **33**(12) (2010) p. 856.

123. W.J. Yang and J.H. Wang, *Proc. 2nd Int. Conf. Mech. Med. Biol.* (Osaka, Japan, 5-7 June, 1980) p. 142.

124. W.J. Yang and J.H. Wang, *A.S.M.E. J. Biomech. Eng.* **105** (1983) p. 263.

125. S.Y. Yeo, *Med. Biol. Engg. Comput.* **47** (2009) p. 313.

126. A.P. Yoganathan, H.H. Reamer, W.H. Corcoran, E.C. Harrison, I.A. Shulman and W. Parnassus, *Artif. Organs* **5** (1981) p. 6.

127. D.F. Young, *J. Eng. Ind. Trans. A.S.M.E.* **90** (1965) p. 248.

128. K.L. Zierler, in *Handbook of Physiological Society, Vol. 1*, ed. W.F. Hamilton and F. Dow (American Physiological Society, Washington, U.S.A., 1962) p. 585.

INDEX